BEACH RAMBLES IN SEARCH OF SEASIDE PEBBLES AND CRYSTALS

LECTOR HOUSE PUBLIC DOMAIN WORKS

This book is a result of an effort made by Lector House towards making a contribution to the preservation and repair of original classic literature. The original text is in the public domain in the United States of America, and possibly other countries depending upon their specific copyright laws.

In an attempt to preserve, improve and recreate the original content, certain conventional norms with regard to typographical mistakes, hyphenations, punctuations and/or other related subject matters, have been corrected upon our consideration. However, few such imperfections might not have been rectified as they were inherited and preserved from the original content to maintain the authenticity and construct, relevant to the work. The work might contain a few prior copyright references as well as page references which have been retained, wherever considered relevant to the part of the construct. We believe that this work holds historical, cultural and/or intellectual importance in the literary works community, therefore despite the oddities, we accounted the work for print as a part of our continuing effort towards preservation of literary work and our contribution towards the development of the society as a whole, driven by our beliefs.

We are grateful to our readers for putting their faith in us and accepting our imperfections with regard to preservation of the historical content. We shall strive hard to meet up to the expectations to improve further to provide an enriching reading experience.

Though, we conduct extensive research in ascertaining the status of copyright before redeveloping a version of the content, in rare cases, a classic work might be incorrectly marked as not-in-copyright. In such cases, if you are the copyright holder, then kindly contact us or write to us, and we shall get back to you with an immediate course of action.

HAPPY READING!

BEACH RAMBLES IN SEARCH OF SEASIDE PEBBLES AND CRYSTALS

JOHN GEORGE FRANCIS

ISBN: 978-93-90294-79-4

Published: 1859

LECTOR HOUSE LLP
E-MAIL: lectorpublishing@gmail.com

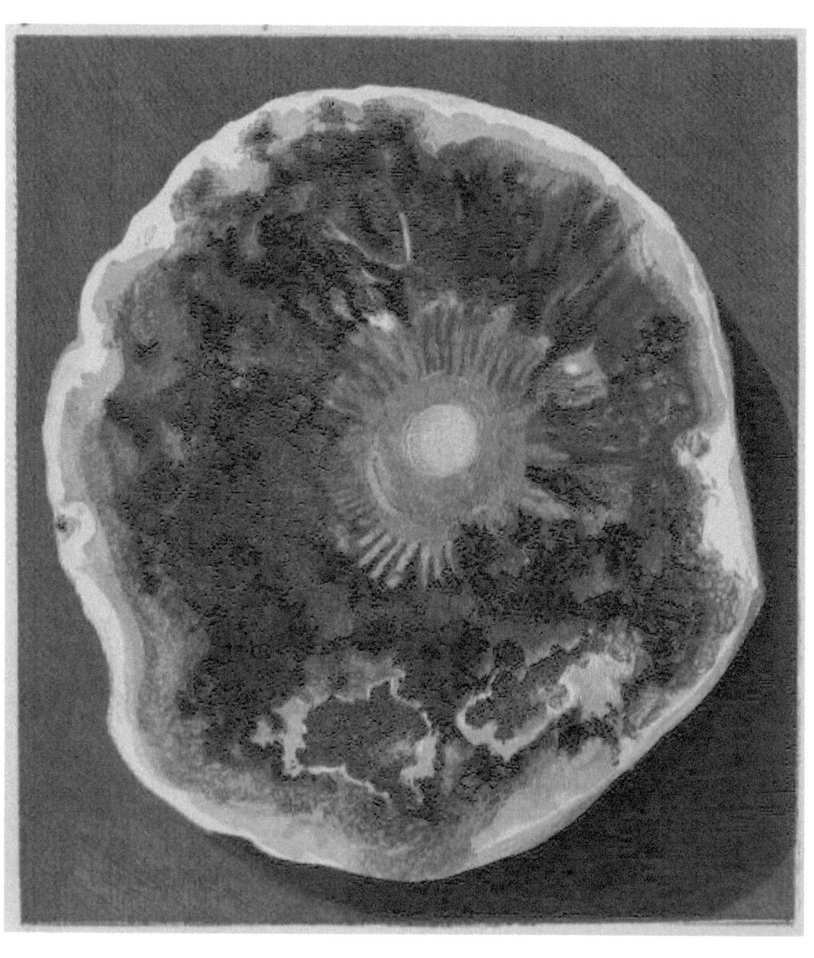

BEACH RAMBLES IN SEARCH OF SEASIDE PEBBLES AND CRYSTALS.

WITH SOME OBSERVATIONS ON THE ORIGIN OF THE DIAMOND AND OTHER PRECIOUS STONES.

BY

J. G. FRANCIS.

1859.

CONTENTS

INTRODUCTION.

There is a pleasure to an intelligent mind in discovering the origin, or tracing the past history, of any natural object as revealed in its structure and growth. It is thus that the study of trees and plants, ferns and field flowers, occupies and delights us. And a similar interest would be found to attach to Seaside Pebbles, as one branch of mineralogy, if we could once come to observe and understand them.

But while the marine shells of England have been all numbered and classified, and even the seaweeds are emerging out of dim confusion into the order of botanical arrangement, there is no popular work extant on the subject of our pebbles.

Dr. Mantell, indeed, published an elegant little volume, entitled "Thoughts on a Pebble;" but he therein treats of a single species, the *Choanite*; whereas, we have other fossil creatures beside Choanites preserved in the heart of siliceous pebbles; and our shores yield from time to time varieties of agate and jasper, differing from the oriental, and some of them of great beauty.

In the present treatise, an attempt has been made to commend this subject to more general attention, by grouping together many scattered facts and methodizing the results. Learned disquisitions and technical terms have been as much as possible avoided; but in the concluding chapters, sundry interesting points in natural philosophy bearing upon the subject are handled rather more scientifically; and here, some original matter will be found.

The coloured plates are after drawings by a well-known and ingenious artist;[1] the original specimens being in my own collection. They have been carefully and faithfully executed, and are on the same scale as the pebbles themselves.

If this essay of mine should induce any one possessed of ampler leisure and more adequate powers to enter more largely upon the merits of the theme, I shall be indeed gratified.

J. G. FRANCIS.

Isle of Wight, 1859.

---

[1] M. W. S. Coleman.

BEACH RAMBLES, ETC., ETC.

CHAPTER I.

Aspect Of A Beach, And Its Probable Origin.—True Nature Of The
Pebbles Which Compose It.

I know of few things more pleasant than to ramble for a mile along one of our southern beaches in the early days of autumn. We get the sniff of the sea-breeze; we see prismatic colours dappling the water, or curiously reflected from capes of wet sand; solemn, beetling cliffs, broken here and there by a green slope, rise on one side of us; while, on the other, we are enchanted by the wild music of the waves, as they dash noisily upon the shingle at our feet, and then trickle back with faint, lisping murmurs into the azure gulf.

Alpine scenery is majestic, and river-lit landscapes are delicious; but they seem as pictures of still life compared with the stir and resonance of the shore and the ocean. The breeze which bends the standing corn does not impart so much pleasure as that which dimples the bay at the foot of our rustic garden; the thunder-cloud resting on a mountain is not so impressive as that huge wall of inky blackness, which seems as if it would choke the very light and air while it gathers on the horizon, but will presently rend asunder and purify the overcharged atmosphere by launching a tempest upon the face of the deep.

There are few persons who, after spending one or two consecutive summers at Eastbourne, or in the Isle of Wight, can repress an ardent longing to visit similar scenes from time to time.

The sea-side stroll has, however, been accused of monotony. But this is either by really incompetent judges, or by inveterate sportsmen, to whom the neighbourhood of the ocean suggests nothing more apposite than a meet with harriers on the downs, or a raid upon puzzled rabbits in some outlying warren, with the aid of a keeper, ferrets, and "varmint" dogs. To such, even a brief sojourn on the simple-featured coast may, undoubtedly, prove wearisome; but the fault rests with themselves. For, all the while, others, who are better informed and more awake to what lies around them, will be cheerfully occupied in kindred pursuits at the foot of the cliff, or away on the beach, or far out, at low tide, among the weedy rocks and sand. Here they hunt the cockle and the razor-shell, collect bright algæ and marvellous zoophytes, or search for agates and fossils among the endless heaps of shingle.

The delicate actiniæ and the rarer sea-weeds cannot be obtained in winter; but the pebbles, which are intended to form the subject of this little book, may always be met with; and the changes induced by rough winds and surging tides, yield them in even greater abundance.

The pleasure of collecting pebbles has been greatly enhanced, to my mind, by considering how it is that we come to have pebbly beaches at all. Inevitable as these may appear to some people, they are quite a phenomenon in their way, and to the full as deserving our attention as the colony in a rabbit-warren.

Originally, the land alone possesses such materials; but it is the sea which finds them out; and these two facts, put side by side, have sometimes reminded me of the arbitrary allotment of the sexes in the old mythology. *Oceanus* being an enterprising gentleman. *Terra* (always feminine) is the quiescent lady, to whom he pays his court. She carries a prodigious number of these treasures in her flinty bosom; but it is only he, and his friends the rivers, who can get at them and draw them forth. In our English Channel, Ocean is as fond of doing this, and of fringing his waterline with a brown pebbly border, as other gentlemen are, now-a-days, of wearing, if possible, a beard like that of the Sophi. Nor is this surprising, when we remember that all creatures naturally desire something which they have not got. For the bottom of the sea itself is no beach at all, but chalk or sand, and sometimes hardened sandstone, with, I dare say, many precipitous pits and hollows, and many pointed crags. Here, gigantic fronds of the oar-weed wave to and fro among thousands of acres of dulse and bladderwrack; while porpoises and dog-fish dive, and limpets and mussels crawl, and arrowy lobsters shoot through the cerulean gloom, and (if Mr. Tennyson may be believed) mermen and mermaids play at hide-and-seek. Wonderful things there must be, if we could only spy them out; but I do not think many pebbles. Whereas, our mother Earth teems with these latter. There are jaspers in the conglomerate, and agates in the mountain rocks, and veins of porphyry and serpentine in the trap and basalt, and garnets and sapphires in the granite, and flint-nodules in the chalk, and quartz-crystals almost everywhere. Probably these exist, also, in many of the submarine strata; but unless a volcanic eruption should occur, there is no force in operation there to dislodge them. The bed of the ocean, and all depths of it below a hundred feet perpendicular,[2] as *divers* well know, remain calm and still, even when a tempest is raging above. But on what we are pleased to call "terra firma," the case is reversed. Solid as the ground appears, all is subject to elemental change and motion; and whenever the waves of the sea or the strong current of a river can plough some crumbling chine, or wear away the face of a cliff, down come the imbedded pebbles and crystals, and gradually form a *beach* upon the margin of the ocean. And this beach is tossed up and down, and rolled to and fro, until most of the stones composing it have become as smooth as hazel-nuts.

The above may be rather a rough sketch of the source of a beach; but I believe it is correct in its leading features. In a subsequent chapter we may better note certain peculiarities which are more than meet the eye.

But what are the pebbles themselves?

---

[2] On the banks of Newfoundland, and to the south of the Cape of Good Hope, extraordinary exceptions occur to this rule; the sea being there agitated to a vast depth, perhaps as much as five hundred feet. But this is probably connected with the current of the great gulf-stream.

Most persons have occasionally handled specimens of the precious stones or "gems;" but few of them, perhaps, are aware that our pebbles of the road-side and sea-shore claim a common origin with these dazzling crystals. Such is, however, the fact. Chemical analysis, availing itself of the blow-pipe, the solvent acids, and the voltaic battery, has succeeded in determining the base of every known gem. And the earths which furnish such bases are chiefly two — ALUMINA, or clay, and SILICA, or pure flint. From these, with an admixture of lime, and sometimes of iron, in small quantities, all the native gems are derived, with the exception of the diamond, whose base is CARBON.

Intensely hard as these substances are, and apparently not susceptible of change if left to themselves, they have probably passed through great chemical changes in the silent laboratory of Nature. For it is supposed that our operations in analysis, if carefully conducted, merely bring back their subjects, by a kind of *reversing* process, to their several primitive bases.

Now it is evident that the commoner pebbles are derived from these same earths, of clay or flint, albeit in a debased condition. For there is nothing else of which they could be made: neither do they exhibit any properties foreign to those which such substances possess. Yet, how vast is the difference between an oriental gem and the brightest production of clay-pits or granite rocks. Not greater, however, than that between Damascus steel and coarse pig-iron; or, between French lawn and sail-cloth. And if Art can work such distinctions, why not Nature?

In fact, a perfect gem is a master-piece which chemistry and crystallization have combined to elaborate, and which man has ransacked the corners of the earth to obtain. The deep has been made to surrender its treasures to the diver or the sunken net: the rock has been blasted, and its inmost vein searched: the rushing river filtered, and its sand sifted: and the contents of the jeweller's caskets are the result. Here may be seen diamonds from Golconda; and rubies from Samarcand and Pegu: sapphires, emeralds, and amethysts from Persia, Arabia, and Armenia: the topaz, blazing like fire, from the Indies or from the Ural: the turquoise, with its delicious blue, from Arabia and Palestine: the opal from Honduras: the blood-red garnet and amandine from Bohemia, Ceylon, and Greenland: the jacinth and the chrysoprase, the chrysolite and beryl, with pure pearls of globular form, all from the land of the rising sun, while a deep brown jasper, rayed with stripes and rings as black as jet, has travelled like a wandering palmer from sultry Egypt or the terrible desert of Sinai.

To form and perfect the finer crystals, extremes of heat or cold appear to be necessary: whereas our clime has perhaps always been temperate. Beside this, it is probable that the mother-earth is not found pure with us. There is a kind of white clay, called "kaolin," obtained in one particular quarry near Meissen in Saxony, and met with nowhere else. From this clay the exquisite Dresden porcelain is baked, and this clay cannot be exactly imitated. A near approach to it has been made by mixing good potter's clay with pounded chalcedony-flints; but still the *biscuit* so produced is never equal to that from the "kaolin." In like manner, we may suppose that the peculiar earth which exists in sapphires no longer occurs in our post-tertiary clay-beds.

Moreover, we know that there are many different clays occurring in our geological strata. We have the clays of the Lower Tertiary; the clay of the Wealden; and the Kimmeridge and Oxford clays, both of which belong to the Oolite. Also in these main divisions, sundry mineralogical varieties are comprised under a common name. But the great age of the granite formation renders it certain that from none of these clays could those sapphires have come (as to their *base*), which are born of the granite rock. Indeed, our existing beds of clay are more or less mingled with felspar, and felspar is one of the "silicates;" whereas the blue sapphire is pure alumina free from all admixture of silica.

However this may be, the great fields for gems are in India and the island of Ceylon, and in certain parts of the Russian dominions. No very valuable stones have as yet been obtained from Australia; although the vicinity of gold-mines has always been held to be prolific in at least one kind: the "mother of ruby" being a roseate substance embedded in the rock, and generally met with alongside of a vein of gold. The topazes are not equal to those from Saxony.

As to our own sea-girt Isle, it is surely as guiltless of indigenous gems, as of white elephants or birds of paradise. Had any such existed with us, they must long ere this have been brought to light and appeared in the market. We have bored the plain to two hundred fathoms' depth: we have pierced the hillside in tunnels which extend for miles: geologists and antiquarians have delved and hammered and sifted: many curious fossils have turned up, and a world's wealth in minerals, but never anything like a diamond or an oriental sapphire.

It is well that, to console us under such apparent poverty as to the gems, we possess the treasure an hundred-fold in other shapes, though derived from the same sources. Clay gives us no sapphires; but it floors our ponds and canals, furnishes our earthenware, and yields the bricks which have built the ribs of London. Carbon refuses to flash upon us in the rays of an indigenous "brilliant," but it feeds our furnaces, propels our steamers and locomotives, and cheers a million of household hearths under the well-known form of Coal. And Iron is our national sceptre: it reddens here no jacinth or ruby; but it supplies us with spades and ploughshares, lays down thousands of miles of railway, and has made England the forge and workshop of the known world for giant engines and massive machinery.

If our wealth be less dazzling than that from Golconda or Peru, it is, we may hope, more durable; flowing to us through a healthier channel, by the honest labour and steady perseverance of the sons of the soil.

This is somewhat of a digression from the subject of Sea-side pebbles. But then, as was said, the magnificent crystals are their near kindred; and in society the custom is to bring in any great connections we may have, on the first fair opportunity: that once done, our *respectability* is supposed to be established.

CHAPTER II.

The Lapidary's Board; And His Workshop.

In some of our provincial towns along the coast, the open door and cheerful bow-window of the lapidary, generally situated in the best street, form a *coup d'œil* which can hardly fail of enticing a visitor to look in once during his morning walk.

If he should do this, he will probably become aware, by a certain whirring sound, that there is an inner room which serves for a workshop. This latter I have always been partial to: but as the contents of the show-room are the most attractive, I will speak of them first, reserving to the end of the chapter a description of the lapidary's wheel and other implements of his trade.

Let us suppose that we are in the pretty town of Sidmouth in South Devon. Somebody wishes for a jet bead to replace one missing from a bracelet, and we sally forth in quest of a jeweller's shop, but come first upon that of his cousin the lapidary, which may probably do as well. Entering the doorway, a gigantic "snakestone" from Whitby, flanks the threshold on one side; and on the other a lump of iron ore, of some two hundred weight, keeps company with a quartz agate of equally cyclopean dimensions. These are striking objects: and instinctively we pause for a moment and consider whence they came. One of them has been washed out of the ribs of the conglomerate: another, after being dug from the bowels of the earth, was sent aloft by the miners as lumber: the third, once the shell of a living ammonite, must have lain for thousands of years in its cemetery of limestone rock, and was only disinterred when some northern contractor, reckless probably of fossil remains, but wide-awake to the actualities of his own generation, was excavating for a railway tunnel. Inland productions these, for the most part. But the threshold is only introductory: pass a few steps onward, and we shall handle substances which are as strictly marine as "crassicornis" or tangle.

The interior of the shop is fitted up with a massive semicircular dresser of elm or maple, some four feet in height, and perhaps half as many in width. This is heaped with specimens culled from various beaches, and several convenient shelves are similarly adorned; the polished stones lying in open trays, but set at an angle so as best to reflect the light. The eyeless head of a Saurian, a creature belonging to an extinct race, is suspended from the ceiling; and a stuffed cormorant, a well-known sea-bird of our own day, mounted on a rude tripod of fir-bough, fills the only spare niche in the apartment. But there is no study or affectation in all this: it is as genuine as the tent of shipwrecked Crusoe. The lapidary has from the first felt himself at home with Nature, and has found room for many of her devices and eccentricities, which he could not now bear to turn out of doors. Neither are we inclined to quarrel with him about his arrangements, though his shop exhibits nothing which will remind us of a frontage in Pall Mall or Bond Street. Moreover, when we look a little closer, he is not such a mere dreamer after all. Commerce has not been forgotten, nor is a certain kind of elegance lacking. Those well-washed panes of crown-glass are decorated with wisps of dried sea-weed more delicate than ostrich-feathers, and which serve the purpose of a hygrometer. And, interspersed with these, are sundry nuggets of amber, bones of the cuttle-fish for your pounce-box, and a string of veritable jet-beads: from which latter we at once select our purchase.

But we are now standing before the central counter, and our attention is drawn to the curious and beautiful fossils which lie upon it. We take up one to which the late Dr. Mantell kindly gave the name of a "choanite." In doing this, he not only adopted a foundling, but conferred endless benefit upon the lapidaries. Nothing will sell in this country without *a name*: the appellation chosen by the Doctor was judged suitable, and a large and increasing sale of the fossil has followed upon it. This one is of a portly size; and the lapidary, after slicing it in two, has polished one of the flat surfaces. The internal structure revealed by this section is not unlike the corolla of a daisy, and at once reminds us of the living zoophyte called "actinia bellis." *Choanite* means "funnel-body:" and the creature which lies here petrified must, when alive, have been globular or pyriform, with many tubular arms branching out from one central trunk.

The petrifaction has been faithful to its prototype: the several tubes being now charged with limestone, and the space between them, once a gelatinous substance, still retaining that appearance in a medium of semi-pellucid chalcedony.

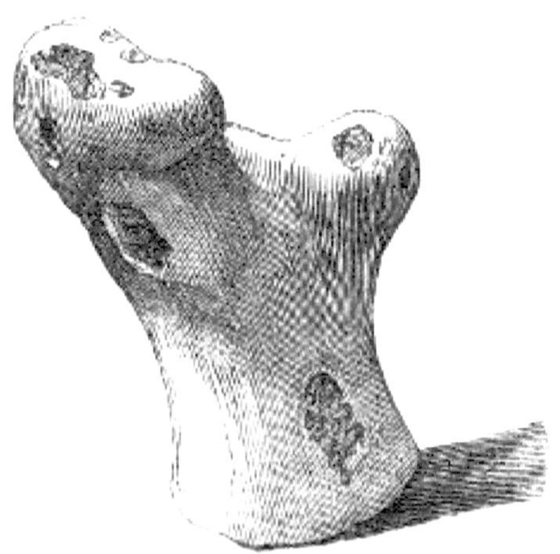

ALCYONIUM DIGITATUM.

By the side of the choanite is another fossil, which we now call an "alcyonite:" the learned name of the nearest living species being "alcyonium digitatum." It is known in the Isle of Wight as "deadman's fingers." Despite the above unpleasing nickname, this is a most beautiful fossil. Its outer form resembles that of a branching ice-plant: while a polished section of one of the stems shows filaments all lying in the direction of the axis, and exhibiting in their cut ends an effect not unlike that of the granulations in a slice from a fresh cucumber. When the pith, as sometimes happens, is particoloured, I do not know a more desirable stone for the cabinet.

Fine specimens of this are now very scarce.

Another of these alcyonites has been polished all round, instead of dividing it: and the pebble being translucent, one can almost count the fibres or tubes disposed lengthwise.

Then we have a zoophyte, not *injected* as are the choanites, but preserved bodily, in delicate gray flint. It is an undoubted "actinia:" in every respect the same with those pulpy individuals who are displaying their jelly-like bodies and floral hues in many a household aquarium. This creature once floated up and down in shallow marine pools, or clung to banks of ribbon-weed fringing the coast-skerries. At present, himself of stone, he is firmly wedged in a hollow within a large pebble, and reminds us of the words of a pretty song:—

"I dreamt I dwelt in marble halls."

Several silicified sponges may next be examined. These are not exactly beautiful, but they are curiously intricate; the once elastic tubes having stiffened into "silex" of a light brown colour, and a horizontal section displaying their fine reticulations.

A step further, and what is that handsome stone? It is a block of Devonian jasper: scarcely so hard as the Egyptian, but more attractive, owing to the richness of its hues. It is striated with veins of agate, and here and there "shot" with a dark metallic "moss" of the colour of bloodstone. As a set off to this, the lapidary has arranged with true taste some quaint pieces of conglomerate, in which the nodules of chalcedony or crystals of quartz have taken a high polish, and are agreeably relieved by streaks of red jasper and blotches of yellow limestone.

Another part of the board exhibits a score of the well-known "echini," which children call "sea-eggs." These embrace two varieties, "spatangus" and "cor anguinum;" or, in the vernacular dialect, "fairy-loaves" and "fairy-hearts." A third form of the Spatangus kind, but with a pointed cusp, is called "Galerites albogalerus." A few are chalcedonized: in others, a bright spar has filled up the interior of the hollow conical shell.

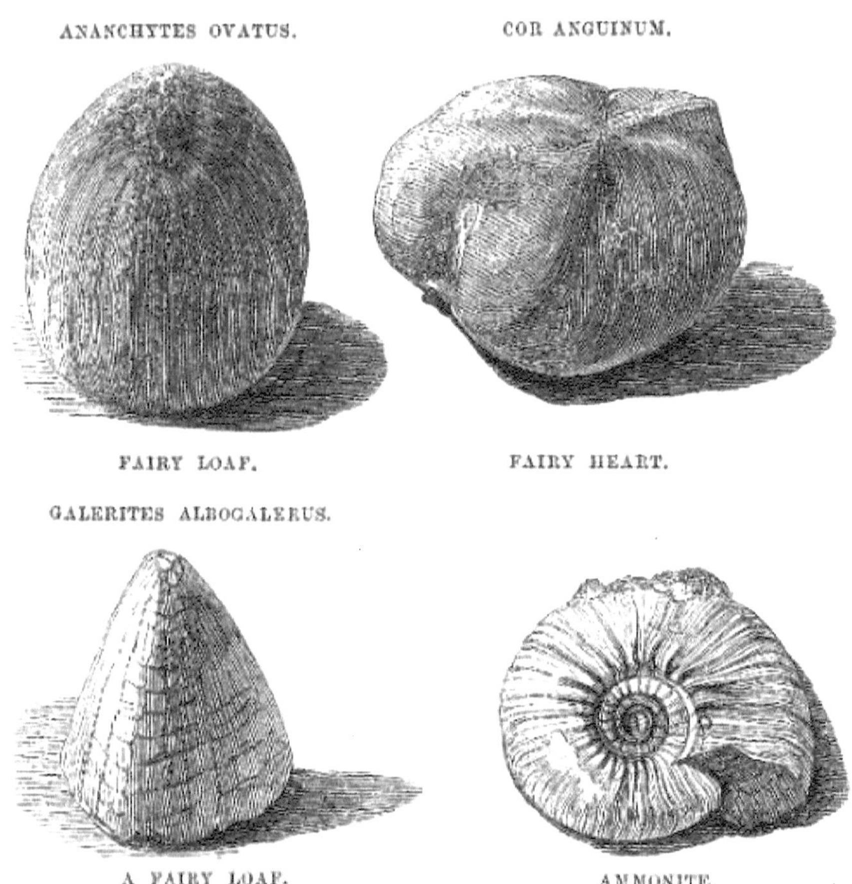

ANANCHYTES OVATUS. FAIRY LOAF.
COR ANGUINUM. FAIRY HEART.
GALERITES ALBOGALERUS. A FAIRY LOAF.
AMMONITE.

A small open tray of twisted card contains several bits of carnelian, and a true bloodstone from the Yorkshire coast: while on a strip of canvas as a *lit de repos*, lie a cray-fish and some belemnites from the Gault formation of the Wight, hemmed round by lustrous sharks' teeth: the latter now hardened into stone, and proclaiming by their isolated condition that the jaws in which they once grew were cartilaginous, and have therefore perished.

Thus the store of the lapidary comprises two main divisions, stones *organic* and *inorganic*, of the past history of the latter we know little; and can only judge by surmise, upon approved geological principles. In the former, there occur plain indications that we are not handling either an accidental "lusus" or an embryo; but that the structure here perpetuated was once endowed with life, and belonged to a creature having its assigned place in the scale of animated being.

Let any one fill a drawer with such specimens gathered by himself: and, omitting the question of mere marketable value, what jeweller's counter can compare with it for real interest? The oriental gems, though they gleam as if "all air and fire," are but dead crystals, and have never stood higher than they stand now: whereas these pretty fossils from our wave-beaten coast tell each a wondrous tale, and form a kind of tangible link between the zoophyte of to-day and his far-removed ancestry in the earliest seas which washed the surface of our globe.

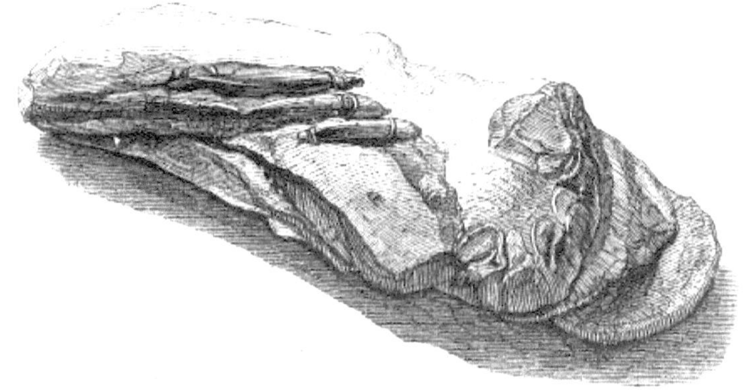

A CRAY-FISH, FROM THE GAULT.

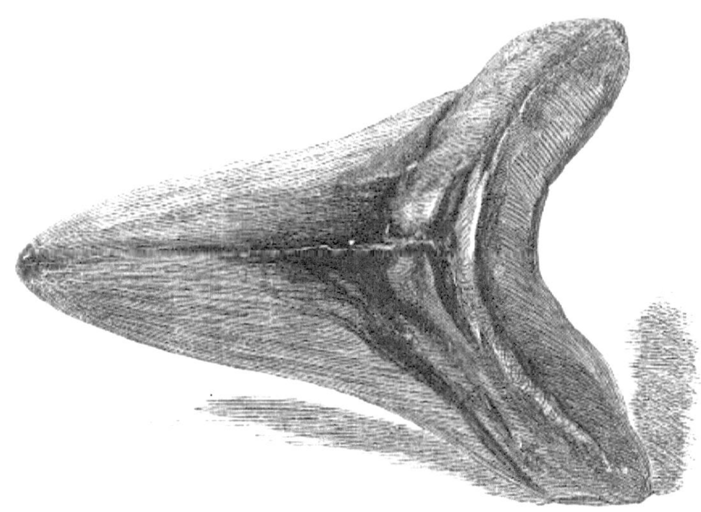

A SHARK'S TOOTH.

A CRAY-FISH, FROM THE GAULT.
A SHARK'S TOOTH.

But let us now look into the back-parlour, where all the cutting and polishing takes place. A peep behind the scenes is generally instructive.

The first time I ever entered such a "sanctum," I remained there nearly an hour; *asking questions*, you may be sure.

The lapidary works by means of wheels. These wheels are instruments for sawing and filing the stones which come under his hand: he has nothing to add to them, but a good deal to take away from them. When any one divides an orange with a sharp knife *equatorially*, he treats it as a choanite is treated in the simplest case: but the principle is somewhat different. The knife may cut the first as smooth as cheese, owing to the superior hardness of the steel and the compressibility of the fruit: but no instrument that man can invent will ever do this with a substance like flint or jasper. All that the wheel can do is to act like a saw, and to take out from the centre of the pebble, in the shape of saw-dust, a section equal in thickness to the wheel itself. If this be effected, and no more, then that part of the work is perfectly done: and this is not effected, without some fault, once in a dozen times, unless the workman be both skilful and attentive. The usual fault is, that the plane of section is not kept perfectly horizontal.

But let us now see how the wheels act. There must always be two of them: one which is to cut by its edge, and another which is to polish by its surface. But the lapidary is generally provided with several of both kinds; partly in case of accident, partly on account of variety in the texture of the pebbles brought to him. Each wheel is fixed upon an axis of its own, of perhaps a cubit's length. And this portable axis is bevelled as a screw, for some inches, at the other end, so that it can be screwed at pleasure into a solid revolving spindle planted upright in the floor of the workshop. The spindle itself may either be worked by a leathern strap and a treddle; or, as is more usual, by a winch-handle acting at a mechanical advantage.

The horizontal wheel being made to spin round by turning the winch with one hand, the workman grasps the pebble firmly in the other, and presses it against the edge of the disk, which revolving in a horizontal plane acts like a saw. A pencil-line may first be drawn on the stone, to mark out the intended section.

But this is not all. A *dry* wheel, although it were formed of the finest steel, could barely scratch the surface of an agate: or, if great force were exerted at the winch, would splinter it. The lapidary has need of diamond-powder, emery, and rotten-stone. He makes use also of a peculiar kind of oil, and has a jar of water within reach for ordinary purposes. The oil is called "brick-oil." It comes from coal-tar, and does not heat by friction, as common oil would: consequently, it neither burns the operator's fingers, nor injures his specimens by causing the wheel to glow too fiercely. Of the above materials, the most expensive is the diamond. Although obtained by crushing "bort," of little intrinsic value, it never costs the lapidary less than twenty-five shillings per *carat*; and he can do nothing without it. Emery, on the other hand, which is a coarse variety of corundum, is cheap enough.

The first wheel put on the spindle is an extremely thin one, it can scarcely be too thin, and is made of tin or of the *softest* steel. Its disk should lie exactly level in the plane of the horizon, and it must not have the shadow of a notch or bend upon its delicate edge. This wheel the lapidary wets, along its entire rim, with oil or water, by means of a feather. A small portion of the diamond, ready mixed with oil, is then applied to the edge of the wheel, the latter being made to revolve gently, until it has taken up the mixture from his finger.

The wheel thus primed is now set in motion, at first slowly, but by degrees more rapidly, and the pebble being steadily pressed, not *pushed*, against it, the diamond eats into the metal, and the metallic edge armed with an adamantine tooth eats into the solid stone, and at length saws it asunder.

In this way an agate of a couple of inches diameter, and of average hardness, will be neatly divided in less than a quarter of an hour.

The two surfaces thus obtained are then inspected, and if the stone be judged worthy, the lapidary proceeds to *polish* either or both of them. There is always some difference between the two, a slice having been taken out of the pebble answering to the thickness of the wheel; hence the importance of using a delicate disk, especially for valuable agates.

To accomplish the polishing, the thin wheel of steel is now exchanged for a much thicker one of lead, which takes its place upon the spindle. It is no longer the edge, but the ribbed upper surface which is now to do the work; and to charge it for this, it is smeared over with rotten-stone dipped in water. This heavy wheel is then set spinning at a rapid rate, and the pebble is pressed flat upon it with great force; the heavier the hand the better. As it slidders about, the action of the rotten-stone, which is very fine in the grain, gives it by degrees a high polish. And this polish is durable, and will preserve the agate from the corroding effect of our atmosphere.

Less than the above will not dress a pebble for the cabinet. Occasionally, much more is needed. For a very fine jasper I have seen two or three polishing wheels employed, the last disk being always loaded with soft rotten-stone.

As to the expense, it will cost from sixpence to ninepence to cut a small pebble through; and to polish one surface, perhaps as much more. This is supposing that you take your specimens direct to a working lapidary. If, instead of doing so, you leave them at a jeweller's shop with directions *to get them done,* you may expect to be charged fully twice as much. Nor is this so unreasonable as it sounds; for the man so deputed acts as your servant and will look to be paid for his trouble: he must have recourse to the lapidary, and you might at once have done so yourself.

To dress a large pebble, especially if a difficult stone in the grain, is a more costly affair. It demands many consecutive hours of labour and unremitting attention, and involves a perceptible outlay in diamond. Moreover, the weight of such a stone causes, by its *vis inertiæ*, a severe strain on the machinery, which then fares like an engine drawing a monster-train up the inclined plane. This wear and tear of the wheels is so serious a consideration in provincial towns, where the supply of "plant" is limited, that few lapidaries out of London will undertake, for any reasonable sum, to dress the agates and close-grained jaspers when they run large.

One of my jaspers, a Devonshire beauty, took eight hours of cutting and polishing in Clerkenwell; and even this was a trifle compared to some of which I have heard tell. Certain substances again are intrinsically teasing to the wheel. There is a class of jaspery flints which have a dodge or twist in their texture; the lapidary abhors these visitors, and will not meddle with them, if he knows it, without bargaining for extra pay.

I have a great pleasure in seeing fine pebbles of my own polished. You can stop the wheel every now and then, and watch how the stone gets on. When the chiaroscuro begins to come out on the coloured pattern, the effect is like that produced by holding some lively object before a mirror. The surface no longer appears flat; but you obtain aërial perspective, as in a good painting.

It were vain to deny that the lapidary's acquaintance might in time prove an expensive amusement; for pebble-hunting is a hobby, and like all hobbies is liable to be over-ridden. But experience begets caution. For a score of stones which a tyro will leave on the board to be cut, a connoisseur will not venture above two or three.

Beginners, however, always run some risk, being naturally enthusiastic; and the best way for them is to lay down a few sound rules, and to adhere to these strictly. One good maxim is, to pay for all work upon the spot. Even an enthusiast will soon grow weary of parting with ready money for mere trash.

Some persons set up a wheel, &c., of their own, and operate upon their treasure at home. I do not recommend this course to any one, unless he were the son of a lapidary, as Achilles was "son of Peleus," and intend to devote himself to the occupation. For most amateurs it will be found difficult. It is five to one that a young hand signalizes his apprenticeship by spoiling his best agates, cutting his fingers, and damaging his machinery. The only advantage, as far as I know, which such a plan may possess, is that you might try some curious experiments in working at odd pebbles.

A lapidary's implements, if complete as they ought to be, will cost him from seven to ten pounds. About five pounds' worth of diamond is a very good commencing stock. Five pounds more will fit up his shop with a counter and drawers to lock, and his work-room with a table, stool, hammers, a few cloths, and a good lens. This is all that he requires, besides knowledge and patience.

Suppose the entire outlay should amount to twenty pounds; in a good watering-place, on the south coast of England, he ought to make from a hundred to a hundred and fifty pounds a year, without doing any night-work. But by day he must not spare hands nor eyes: and he cannot afford to make any serious mistakes with his customers. If he cuts a fine choanite the wrong way, or mutilates a rare alcyonite, or robs a promising "landscape-pebble" of its pellucid sky, it is well-nigh over with his professional reputation; and that once gone, he will get little custom of value. When I was last in Ventnor, the sea had thrown up some singularly fine pebbles for several successive tides. Three working lapidaries rented houses in the town; but all the more valuable orders went to one of them. This man knew admirably well how to handle almost any stone which was brought to him; while his two brethren in the craft, though both of them hard-working men, were comparatively ignorant. They were like quack-doctors, and he was a learned professor.

CHAPTER III.

The Contents Of A Good Beach, And How To Obtain Them.

I have always been fond of trying conclusions with Nature at first hand. When a boy at school, I learnt to swim by following some bigger boys, who could swim already, into water which was quite out of all our depths.

I had never till then tried, except in shallow places, which, whatever people may think, afford no criterion of how it will fare with you in a drowning current. On this occasion we swam about fifty yards, in ten-foot water, not without trepidation at first on my part, but towards the end I gained breath and confidence. After that, I never had any difficulty, whether in lakes or rivers, and I have ventured into scores of them at all seasons, and that without the slightest previous notice of their depth or temperature.

In like manner, I am convinced that to master anything, whether bodily or intellectual, the all-important point is to make a genuine effort of our own. Books may do something for us; teachers, if we can afford them, something more; but we must do the main part ourselves.

I am reminded of the above in proposing to describe a good beach. A lapidary once said to me, "Sir, you know this bit of the coast as well as I do myself." It was a high compliment, but it was not unmerited. The truth is, I possess "memoranda," jotted down from time to time, of divers beaches and their usual contents, and I find that, omitting all mention of bare reefs in shale and sandstone, or long reaches of sand, which are of continual occurrence, there remain some forty odd miles of shingle lying in three different parts of the English coast, with the character of which I am thoroughly well acquainted.

Some of these beaches are very superior to others, and I select such for description.

Amid countless boulders of flint, and heaps of hardened gravel, we have upon a good beach certain smooth, translucent pebbles, and we have fossil petrifactions enveloped in an opaque crust, and we have the variegated jaspers and moss-agates.

Our semi-pellucid stones were, with the exception of an occasional bit of "bloodstone," the only pebbles of native growth known in Britain half a century ago. They consist of two or three varieties of agate, two of carnelian, and one of the crystal called an aquamarine. The presence of this latter on the coast is a mere accident, as it is for the most part a far-inland production, the growth of granitic rocks.

Agates and carnelians were once of frequent occurrence, but they have now become scarce. Their brilliancy insures their being instantly seen, when not buried under the loose shingle.

AGATE, with us, is always found smooth, and some of the best specimens are indented on the sides, as if they had been subjected to a pinch or pressure while in a soft state. As this substance is not met with under the spiked form of a crystal, we may suppose that it never was crystallized.

It is not laminar, like the diamond, nor coated, like a pearl, but one simple concretion. It contains some alumina, but more silica, and is probably one form of the onyx-stone. It is seldom very bright, being more or less debased with earthy particles, but it is a pretty thing to pick up, and it takes a high polish on the wheel of the lapidary. I believe it is harder than jasper, but perhaps scarcely so *tough*.

In England, we find the agates greenish white, lemon colour, or dark grey; and on the island of Iona, hard by the ruined monastery, they are picked up of a soft green hue, and as clear as a chrysolite. The best lump of agate I ever saw came from an unfrequented bay in the Isle of Wight; the colours were dullish, but in texture it approached an oriental onyx, and it weighed above a pound. I know of no equal specimen in the collection of the British Museum.

CARNELIAN, which is our purest form of chalcedony, is undoubtedly a more beautiful stone than the agate. This is either milk-white or of a deep red. The latter tint is becoming scarce in Britain, having been much sought out for the manufacture of seals and ring-stones, but now that "sards" are so much in vogue, the real carnelian may get a respite and appear in force again, as salmon have sometimes been known to do in their native streams after falling off for many years.

The "sard" may be considered as the carnelian of the desert, or the "carnelian" as the sard of the seashore. Sards are plentiful in the east. Travellers, whose path lies over the waste plains and sandy reaches of Egypt or Lybia, should pick up any darkly-tinted pebble they may descry on the surface of the ground: it is probably a sard, perchance a valuable specimen. One day, in homely Brighton, stepping into a lapidary's, I found upon his counter half-a-dozen unusual-looking stones already cut in two, and some of them polished. "These are not English," said I, "where did you get them?" He told me, in reply, that a gentleman just arrived from Egypt had taken these stones out of the mouth of his carpet-bag and left them with him to be dressed. I examined them closely. As I had expected, on learning the locality, one was a blood-red sard, and two others were jaspers; a fourth was curiously mottled and ribbed with chalcedony. They had been obtained without any expense, and at no trouble beyond that of stooping to pick them up, and all four were very *saleable* articles in the trade.

The finest red carnelians are brought from the East, but they occur also in Silesia, and splendid specimens of a dark hue have been obtained from the sands of the Rhine.

In Scotland, on the beach of St. Andrew's, I found a pretty variety, "eyed," but it is scarce. I have since seen similar stones, which had been picked up at Cromer or at Aldborough.

The AQUAMARINE is sometimes met with on the sea-shore at Aberystwith in North Wales, and more rarely on our eastern coast. Of course it is a crystal, which has come down from the inland rocks, and has afterwards been rolled smooth by the action of the waves among other pebbles.

The base of all our English crystals is *Silica*, with an admixture of lime. Rock-crystal is the purest form of Silica. Common spar is a carbonate of lime. The black or grey flints, which are shed in myriads from many a chalk-cliff, are "Silex" much debased; some dark, viscous matter, such as bitumen, having united, I think *chemically*, with the clearer substance. Quartz are the small crystals of silica; a mass of these will vary in its configuration.

Cairngorum stones are rock-crystals from the mountain of that name, deeply impregnated with iron and aluminium. The claret-coloured ones, found on Ben M'Dhui, are our nearest approach to an amethyst, and some deep-red specimens remind us of the garnet.

As I propose here to speak only of our sea-shore pebbles, I shall not dwell upon the Scotch crystals, which are, moreover, familiar to almost everybody in the ornaments of tartan dresses. Neither shall I touch upon the fluor spars of Derbyshire, or the magnificent crystals found in the mines of Cornwall and Cumberland.

All these belong to crystallography, and I am persuaded that, in our productions of the seaside, a crystallizing process has been the exception and not the rule. I will only add to what is said above, that, as far as I know, amidst the great range of tints comprised in the different crystals from Scotland and from Ireland, there is no instance of the *peach-colour* or delicate pink, such as are picked up on some of the "moraines" on the Alps of Savoy.

Secondly, such parts of our coast as are hemmed in by cretaceous or sandstone cliffs abound in fossil pebbles. The organisms which these inclose are, almost universally, those of zoophytes. Dendritic or vegetable markings are rare, occurring principally in the white chalcedony. A good pebble of this kind used to be called a mocha-stone; the difficulty always was how to be quite sure where it came from. Those from the Indies would generally be handsomer; while, on the other hand, an equally fine specimen from our own shores would be of treble value.

But the petrifactions described in Chapter II. may be readily obtained throughout the entire range of coast from Hastings to Selsea Bill, and further.

In gathering these pebbles, if you intend them for the cabinet, and not merely to amuse a passing hour, regard should be had to three points; the size, the pattern, and the colour. The last-mentioned, which is, nine times out of ten, the cause of average specimens being seen at all, will itself be determined by the material of which the pebble consists. Chalcedony has a bluish cast in the solid stone, especially when it is wet with brine; but where "moss" is present, this will cause a golden or red tint. Jasper, when semi-transparent, is greenish, otherwise, a blood-red.

As to size, choanites and the globular "sponges" will vary from that of a small pippin to that of a full-sized orange. Very rarely they exceed this. If the pebble be of the right sort, the larger the better: but, as a general rule, the finer ones run small.

The pattern is the most important feature; as you will soon recognize in making a collection. And this, moreover, if no injury has yet been done to the stone, decides the "contour" of the rolling beauty.

I shall not, however, say much about these patterns; not wishing to be charged with exaggeration, as I probably should be, by those who have never seen good ones, and who do not conceive of the endless varieties into which four or five colours can thus be wrought in the stony loom of Nature.

Amateurs must learn for themselves what these really are, and how to judge of them scientifically, when they pick up a fossil weighing, perhaps, a pound avoirdupois, and looking like a champion-potato. One or two hints may be ventured. If the creature was an "alcyonite," its *facsimile* in stone should have the external rings or mouths clearly defined, and, if possible, equi-distant. Such a one, when cut, will exhibit its tubes evenly disposed—not unlike a section of fir-wood under the microscope. If it was an "actinia," then the body of the zoophyte should be central in the pebble, and the tentacles will be not much melted away in a good slice.

"Sponges" must be chosen principally for their colour; in other words, for the *texture* of the stone. Those which exhibit the reticulations, white or straw-colour, upon a very dark ground, are the most effective generally: but the most perfect one I ever saw had a blood-red pattern upon white. For a single *slice* from this stone, half a guinea was offered by me, and refused.

"Choanites" are easily discerned.

The "ventriculite" must have been a creature lower down in that scale than the choanite. This fossil is repeatedly met with on our coast, but I do not admire it enough to have retained a single specimen. When alive, it would appear to have resembled, in stature and configuration, an ordinary toadstool.

The choanite was, undoubtedly, a beautiful creature, and, as ten thousand of the family testify, abundant. Dr. Mantell said that its form, when complete, was that of a pear or fig, and I think he obtained such a fossil from the Lewes chalk, where it was growing upright on a stalk, in the way in which he afterwards depicted it. Looking over my collection, I see that I have a similar one, in black chalcedony-flint, which I picked up at Shanklin; only, in my specimen, which came from the beach, the delicate stem is, of course, gone.

The complete pyriform mound is rare, for obvious reasons. I never saw but three instances of it; the above, of my own finding, was one. Another was in pellucid white agate, spotted all over with the ends of the feelers. The third, which I also picked up, had been pounded on a rough beach, and crumbled in my hand.

Choanites and ventriculites, as animals, are supposed to be extinct. Perhaps they are so, though I do not see how any one can take upon him to pronounce as to what living organisms the great Southern Ocean may or may not contain.

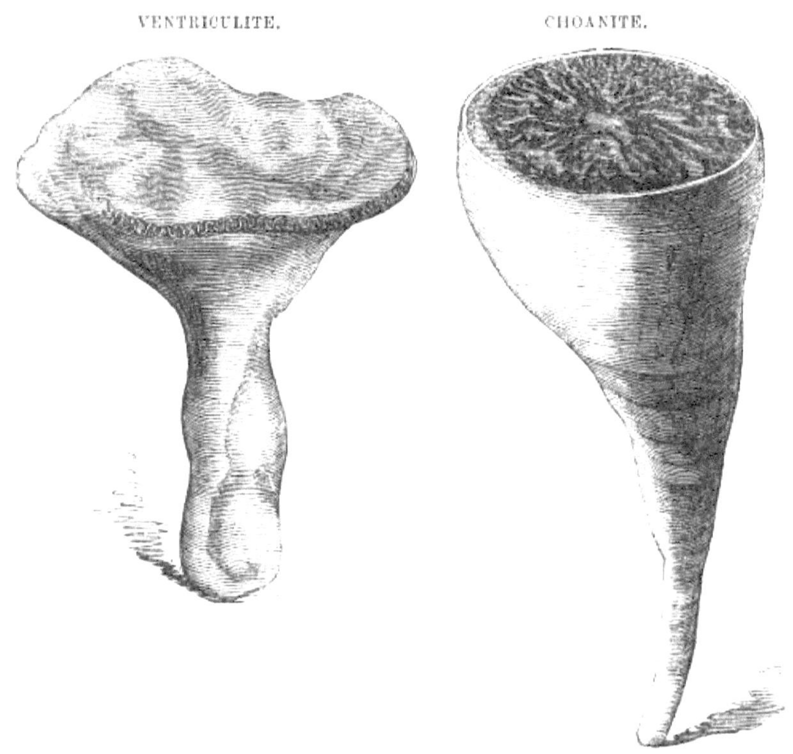

VENTRICULITE. CHOANITE.

The creature, however, which I admire most, as perpetuated in these marine fossils, is not a choanite, but an actinia of the "crass" kind. One of these, large and of a globular form, in which the tubular tentacles are distinctly shown, and the colour is yellow in the agate, I found, in a sequestered spot, where the deep sand must have received and sheltered it shortly after it had dropped from the cliff. The outer whitey-brown crust was unbroken: and as this crust is the cuticle of the pebble, and which always wears away first, I have no doubt of the specimen being a perfect one. It could never have undergone a rub, beyond that from soft sand or softer seaweed.

Lastly, striped jaspers and bloodstones are to be had for the seeking, on our Devonshire and Yorkshire coasts. The bloodstone is too well known to need any description; neither is there much variety or interest in a rush-green, spotted with red. But the jaspers embrace several other colours, and many lively patterns. South Devon has good ones, resembling agatized wood. Scarborough has good ones, some of them quite equal to the "weed-agate" of India. At Eastbourne, a dark-brown variety is occasionally found, which is highly prized, because it approaches the character of the Egyptian. Fine pebbles of this kind are also to be obtained in the neighbourhood of St. Alban's, in Hertfordshire, but they are liable to be gritty.

Beside the above, which are true jaspers, scores of jasper-flints in bright red, yellow, or green, occur on almost any beach. These are simply burnt flints, containing a portion of oxide of iron. Their present condition is probably the result of extinct volcanic agency. These are not nearly so hard as the true jaspers, and their *fracture* is "conchoidal."

For pleasing the eye, perhaps the choicest stone in Britain is a moss-agate, of which the pretty name yields an accurate description. The "moss" is some oxidized metal, whose ramifications form a striking contrast with the limpid chalcedony in which it seems to float like seaweed or sponge.

When a fine specimen of one of these stones has been cleverly cut, it is not unusual to obtain in its section the principal features of a mimic landscape—the clear sky, and the fuscous earth. Some, in addition, display a setting sun, &c. Such stones get the name of "landscape-pebbles."

On the shores of Loch Tay, in Perthshire, remarkable ones are found—in which the imaginative Highlander fancies he can trace the features of his beloved mountain-scenery. Great store is set by these, on account of their pattern; but they are mostly in sepia, and white, being never suffused with warm colours, as are our marine specimens.

I may note here, finally, that to a true connoisseur, the prettiest moss-agates are not so welcome as those which exhibit a bold and haggard style. Some of the French pebbles, from the neighbourhood of Dieppe, would be very fine indeed, only that they are spoiled by an undue quantity of black "moss," as black as mud.

Most of the above-named treasures may be obtained by any one who owns a fair amount of perseverance, provided he or she have an eye for pebbles. And nobody need be discouraged from the search, who likes them well enough to be willing to take a good deal of trouble to find them. One-half of what are counted difficulties in this life have their root and growth in our constitutional laziness, and may be overcome by a little energy. We have all of us an eye for ripe cherries and red roses, why not for pebbles?

Let me now state in a few brief paragraphs the merits of the case. People say there is luck in all things. It may be so, once or twice; I never knew it hold for a continuance. A course of practical experience in almost any department of life will bring a sure quietus to such crude fancies.

The searcher after pebbles must look to more sterling qualifications. He should have good legs, good eyes, good judgment, and—I may as well say it at once—a good temper. He should choose a likely part of an unfrequented beach: and should go down at the right hour, somewhere about half-tide, when the tide *is running out*.

It is as well for him to walk with the sun a little behind him, and on one side; and I would recommend him, if possible, to avoid entertaining a north-easter in his teeth.

Then he should be suitably rigged for the expedition. If he has ever fished salmon, he will need no wrinkles here: but in case not, I may specify the following.

Strong ancle-boots, with double soles and on real hob-nails, and a stout woollen sock within; rough trowsers, which he will not grudge damping in the brine; and a coat of fustian or tweed, with ample pockets. When the season is wintry, add to these a warm wrapper for the neck; and, in one of his pockets, a Cording's india-rubber cape with sleeves. A water-proof cap, with a curtain to it. No umbrella, no stick; but a light geologist's hammer; and a canvas-bag, worked with open meshes, for the heavier specimens, which otherwise he would have to carry in his hand.

If he is going far, and it is as well to count upon eight or ten miles from home, let him carry a flask of any liquid he likes best to imbibe; together with a hunch of bread and cheese. [N.B. This will be found more sustaining than a meat-sandwich.] Lastly, a few mild cigars, not omitting the usual implements for striking a light *sub dio*. The fragrant weed is always a cheerful companion, and doubly welcome when your path lacks flowers: moreover, it has historical associations, which cabbages have not, and will bring to mind Sir Walter Raleigh, who was a great man.

The above rules concerning the toilet will hold good for the fair sex, *mutatis mutandis*. Only in addition, ladies should engage some stalwart arm to lean upon; and should on no account venture themselves among slippery rocks. As for those knights and esquires who accompany the gentle dames on such a quest, they must look to themselves. I cannot now stop to warn them of their peculiar peril; but they know what I mean. It was among the Ocean-isles that Briseis led captive the stern son of Peleus.

Above all, let the Pebble-searcher have nothing on his mind when he sets forth: no broken engagements, no crying debts. Otherwise, he may look in vain for moss-agates. Shakspeare, when he has to account for a valiant warrior and politic general, a crowned king to boot, losing the great battle of Bosworth-field, ascribes it all to the ghosts who sat heavy on his soul. Any neglected or injured creature may prove on such occasions a vengeful ghost.

Supposing your hours free, the best are those in which the daylight is most powerful: say, from ten in the morning until four in the afternoon; or, to take things in moderation, from eleven to three. A beaching expedition made thus with determinate purpose, once or twice in the course of six days, I hold to be reasonable diversion; and for matter of interest, I know of no out-door pursuit which excels it, not even Insect-catching, or Fly-fishing.

The above hints being attended to, one important point remains still to be considered. It is absolutely necessary to understand pebbles in the rough.

The more peculiar a beach is in its contents, the less will it exhibit to catch the eyes of one who has never learnt to appreciate them. Coloured seaweeds are enticingly bright; pearly shells are as evident as trinkets; but the external coat of a genuine pebble differs widely from its internal structure, and may yield to the uninitiated no indication whatever of the value of the latter. A crust of hardened lime or sandstone is a frequent envelope of the best specimens in one kind of zoophyte. The stone itself is very likely mis-shapen, and perhaps lies more than half buried in sand or shingle. Without knowing anything of the nature of these fossils, you may occasionally pick up such "darkies" at a venture; but you will never feel assured of them, you will never be able to glean them from the multitudinous gravel, as the determined Fly-fisher gets the best trout out of a pool, if you have not the pleasure of an intimate acquaintance with their species. The readiest way of acquiring this, and it is not born with anybody, is to spend a wakeful half-hour, once and again, at the board of any civil lapidary, and there thoroughly to inspect his casual store. A trifling purchase will abundantly content him in return, the more as he hopes to cut several pebbles for you.

Observe, then the peculiar way in which the outer crust of an old pebble is worn, as compared with that of one which has newly descended upon the beach. This crust or cuticle is highly suggestive, if you can once come to understand it; and as a hopeful indication of progress, you may safely assume that you *do* understand it, when you find that you can obtain pebbles such as nine persons out of ten somehow never "have the luck" to meet with. Sometimes a stone has parted with its original coating, and has donned another: and this operation may either be complete, or still *in fieri*. In the latter case, there will always be some token; either by the substance not lying evenly, or by a change in the colour of the second envelope, which becomes apparent on chipping off a fragment from its surface.

Lose no opportunity of perfecting your judgment upon any form or texture that comes in your way: for upon this will really depend what sort of collection you shall win from the bosom of the coy beach. Nothing is easier, on an average coast, than to pick up a score of showy-looking but inferior stones. Few things will be found more difficult than to bring home, as the fruit of your morning's walk, two or three valuable specimens. And the cause of this lies not at all in luck, nor altogether in mere labour; for in all departments there are dunces who will still drudge hard. It lies in the presence or absence of sound information on the subject.

A good rule is, even early in your apprenticeship, to eschew trumpery. But beware of hasty conclusions. Some collectors will take nothing away with them from the beach which they do not feel quite sure about. This is shallow practice, for it assumes an amount of discernment which nobody possesses. I can see as far into a stone wall in this sense as most people; but I rather like to carry off occasionally an odd-looking stone which, like Bassanio's happy casket, "rather threatens than doth promise aught." Such a pebble may turn out a prize. The things which every one should reject without hesitation are those whose character is evident, and with it their little worth.

Mediocrity in pebbles is insufferable. Do not assume that the *wet* portion of a beach necessarily yields the best chances. Many pebble-hunters never get over this early delusion. On a coast which is rarely trodden, the upper range contains the richest store. And in every case, the tide runs twice in the twenty-four hours over banks which now look dry and dusty, as well as over that narrow strip which you see glistening with saline moisture. And while you cannot fail to remark a pellucid stone or fossil which is sprinkled by the wave, if your attention is caught by anything lying high and dry, the latter is probably worth stooping for.

Walk rapidly over a poor, unproductive beach; but take your time and make good use of your eyes when you find you are in the land of plenty.

If you carry a hammer with you, let it be of the right sort; a well-tempered metal, not too large to stow away in a hind-pocket, and *not armed with a spike*. When you have such a one, it is tolerably safe to strike even a heavy blow on a large pebble in your other hand; but if you do the same by a small one, you may chance to maim your left palm for life: because in this case there is little reaction, and the blow you deal will *drive* the stone. But even with large pebbles, the preferable method is to plant one foot firmly on the stone, leaving a portion exposed to your aim. If your boots are of a sensible thickness, nothing will come to grief.

Never dash pebbles down upon others in order to break them open. In the first place, you risk a wound in your face from the rebound of a fragment in what mathematicians term the "angle of reflexion," an angle which you have probably not calculated; but besides this, you may by such imprudence destroy a valuable specimen. You can open plenty of fossil flints and waterworn jaspers in the manner I have recommended above; which knocks off a corner, and shows you the true nature of the formations you are walking upon. I have learned a good deal in this way.

Lastly, when you are landed upon a superior beach, with the promise of a few hours of open weather, use the golden opportunity and do your best.

Solomon was quite right when he said, "Whatever thy hand findeth to do, do it with thy might." It is evident that if you do not *in foro conscientiæ* think such a hobby worth the while, you ought not to meddle with it. I quite agree that making shoes is more profitable, provided always that you have custom; although I have known more than one instance of working craftsmen who at intervals picked up some very good pebbles too.

The above rules may sound marvellously simple. I only wish that a few as simple were generally acted upon. We might then hope to see some noble collections of really instructive pebbles from our as yet little known sea-coast.

CHAPTER IV.

Which Be The Best Beaches In England.—Concerning Stationary And
Movable Shingles.

The variety exhibited by Nature seems to be almost infinite, take what ground we will for observation and comparison. This has been frequently noted in both the animal and vegetable departments of Creation. I believe it to be no less true of the mineral, especially when we embrace under that head fossil remains.

If there be any feature in British scenery which to a cursory observer appears uniform and identical everywhere, it is the yellow-looking beach along the coast line. Yet it will be found on trial that our beaches differ widely from each other. No two are quite alike. Devonshire is not more diverse from Yorkshire, or Norfolk from the Isle of Wight, than are their respective beaches, as to what these contain.

Nay more: on the range of the Sussex coast—a very monotonous range—I am acquainted, at this present time of writing, with four beaches, which I do not scruple to pronounce totally different one from another. If I were shown the pebbles, I could generally tell from which quarter they came. But to become aware of variety, you must observe certain admitted facts. The frequenters of our coast-scenery, not one-tenth part of whom can plead the sad excuse of being invalids, usually know but little of the character of a beautiful beach lying in their intermediate neighbourhood. Ordinarily speaking, a man will be better informed concerning the plains of the Pampas, or the windings of the Coppermine River, than he is upon the nature of the soil which is under his feet, or of the bough which hangs over his head in the forest. And the reason of this is obvious; we neglect the reality for the sake of the appearance. We greedily devour certain stereotyped works as a kind of royal road to knowledge; but we omit to notice and inquire into the sensible objects which are strewn along our daily path.

Of course, we pay the penalty in kind. Two-thirds of our "book-knowledge" is merely conceited ignorance; not the genuine gold, fit for current coin, but a baser metal washed over, which will stand no one in stead for long.

I will now describe accurately three beaches which occur in three different counties. Any one who makes their acquaintance hereafter, especially if in quest of pebbles, will, I believe, recognize the original of each sketch. One of them now lies far away; but to think of them is like dwelling on the still vivid impressions of a dream.

I am in a district of the New Red Sandstone, and as far as my eye can reach I see towering cliffs of a kindred formation, wheeling round or jutting out in foxy-coloured masses, while the innumerable points of shingle glow like an autumn stubble at their feet. I commence searching near the bright water-line, and what with the simple grandeur of the scene, and what with the balmy breath of the south, I insensibly stray onward for several miles, until the increasing weight in my pockets suggests a pause. Now I will sit down under the shadow of that mid-way rock, light a cigar, and inspect my treasures.

No sooner said than done. What a calming sensation a whiff of good tobacco induces on these burning days! But what have I got? Above thirty globes of chalcedony, blue and white, as oval as bantams' eggs. I select half a dozen on account of some beauty or peculiarity, and fling the remainder into the sea, there to undergo a fresh impregnation. Next, sundry bits of the red and white conglomerate. Let me advise everybody who has the opportunity, to pick this up. When sliced and polished, the surface obtained beats the inlaid work of a cabinet-maker all to nothing. Next, a couple of choanites, one in the dark "rag," ribbed like sea-weed; the other in pretty light-brown moss-agate. These two go into my breast-pocket for safety. Lastly, what has never been out of my hand since I found it—a huge, knotted jasper-agate, of five pounds' weight.

I will only add that the beach here described is a first-rate one in its way, although not my absolute favourite. I had, however, never seen it before, and have never visited it since; moreover, I had not a hammer with me, nor any refreshment beyond a cigar, after tramping over many miles of deep shingle; consequently, I do not suppose I did justice to the contents of that bit of coast.

But where am I now? Two hundred miles away, and the scene is changed. I have just rounded a headland, on whose summit the ripening corn-field is waving and gleaming like the tuft on the head of a woodpecker. Dizzy precipices of chalk stand like sugar-loaves against the blue sky, here and there cleft in a tortuous, yawning chasm. Beneath, the beach seems to slumber, so gentle is that lisping murmur, where the long billowy swell washes the face of the skerries, and lifts up and down their heavy fringe of tangle like a giant's beard. Two hours are mine, as we mortals reckon, before I am expected at the evening meal. This time I have my hammer with me—no clumsy carpenter's tool, which might rend or splinter, but a tempered steel.

I did but walk a couple of miles in all. I found sundry pieces of perfectly white agate, one of which proved good enough for a seal. I broke open a score or two of jaspers, and was rewarded at last by a fine slab of the rich brown (quasi-Egyptian). Then I hunted up from the shingle a pocketful of "sponges," and a choanite in dark purple flint, which is rare. And then time was up, and scaling the cliff by a well-known gap and pathway, I returned home to crab-pie and a cup of tea, and the welcome of a dear, kind face. This part of our coast is much visited now by pebble-seekers.

Now I have crossed a channel of the sea, and have looked upon the mighty three-deckers, and all the flower of the yachtery of England; and I have strolled through a smart modern town, and have passed a meditative half-hour in a very ancient village, and then, at a sharp turn, I have quitted the high-road, with its hot milestones, and picked my way through a blossoming lane where the nightingales were singing; and I have scrambled over the edge of a snipe-bog, and hurried past the mouth of a diluvian estuary, and here I am, stretched at full length on the choicest bit of beach in all the realm of Queen Victoria. In front of me are soaring pinnacles of chalk, ribbed and dotted with primeval flints, like the gun-muzzles in an upper deck. Behind me rises another barrier, formed of the dark "greensand;" and on one side, for the space of a mile, these lofty brows have dipped and disappeared, that the red edge of the wealden might crop out from a wide tract of rush-grown water-meadows. Through the heart of these latter steals a rivulet, beloved of stray wild-fowl, but not tenanted by anything so lively as a trout.

The position of this beach is all that can be desired, and its contents will not disappoint expectation. It has sometimes reminded me of one of those angular nooks formed externally by the converging hedges of a cover and a crop, where you are sure to meet with the hares and pheasants which have come out to feed on your wheat-stubble or turnips. Here, however, instead of the scuttling game, you must look for matchless pebbles.

Often as I have paced this charming strand, the occasion of my first stepping upon it was pure accident, as, indeed, are some of the pleasantest turns in life. I remember one summer's day, when the solid cliffs seemed to be pulverizing under the heat, and the air-vessels in exposed sheaves of bladderwrack were splitting and popping like crackers, I found four or five specimens within a couple of hundred yards which I have never since surpassed, seldom equalled. Two of these were *deep red* choanites, conveying the idea, not of dull "oxide," but that the animal's blood had suffused the agate; another was a particoloured madrepore, another, a moss-jasper.

This bay, however, possesses peculiar advantages. In addition to the fossils which drop from time to time out of its own overhanging cliffs, it is fed in a remarkable manner from other sources.

Until of late years, Sussex has been the usual resort of those who desired to collect pebbles for the cabinet. The capabilities of this coast vary with every twenty miles. The Brighton beach was once famous for the supply of "landscape-pebbles;" but in the lapse of years its fecundity has been sadly impaired. The best chance remaining now is that of an occasional waif during the March winds on the sweep fronting Brunswick Terrace, and for about half a mile toward Shoreham.

Bognor has long reefs of rock at low water, and, by the usual rule, it ought to yield petrified sponges; for I think that, generally speaking, the locality determines the animal. But the infirmity of the Bognor shore is, that its pebbles are chafed and worn to skeletons by the furious storms which assail this unprotected coast in winter. No fossil with a delicate constitution can long withstand the effect of these gales and plunging seas.

Off Beachey-head, in the other direction, is found a variety which I have not met with elsewhere. The specimen shown to me was vermicular; the creature, which had perhaps been a "sabella," lying in distinct coils, not unlike *macaroni*, in the blue agate. This pebble was a very large and perfect one; and the man who had picked it up, and paid perhaps half-a-guinea for the dressing of it, was desirous to realize a five-pound note. I wish much that it had found its way to the British Museum. There is no stone there resembling it.

At Deal and Ramsgate, "fortification-agates" may be readily met with, and sometimes a bit of carnelian: but this last, bright and homogeneous, is rare. The Hastings beach is mainly worn out. Aldborough, in Suffolk, was wont to be famous for its coloured agates; and Fifeshire, and part of Forfarshire, for eyed jaspers: but many years have gone by since I visited either of these counties, and I would not venture to speak with confidence concerning their actual beaches now.

Very good pebbles are picked up at Cromer; and, occasionally, carnelians, both red and white. Felixstowe, a little lower down the coast than Aldborough, abounds in yellow agates, some of them beautifully translucent, and rivalling the carnelian in smoothness of texture. At Scarborough, remarkable specimens both of the agates and jaspers are obtained by those who know where to look for them. Filey has the same charms for enticing strangers to wander onwards along its pretty coast. Indeed, all the Yorkshire beaches deserve and will repay a visit, both in fossils and transparencies.

Carnelians are peculiar stones as to their habits, if I may use such an expression in speaking of inanimate substances. It were absurd to doubt that plenty of them still haunt our English coast; but where they manage to hide themselves has often perplexed me. The method of searching for them is as follows. Instead of keeping the sun behind you, he should now be full in front, like a candle in your eyes. But you must not look at the sun: you must look along the surface of the beach; and whenever a bit of carnelian peeps out from the shingle, you will at once descry it, owing to the position of the great light.

I tried this once at Blackgang Chine for three-quarters of an hour, and in the course of that time I filled both my waistcoat-pockets with red bits. It is true, none of these were larger than pudding-raisins, and some of them scarcely more brilliant; still, the search was very diverting. If the shingle had not been unusually damp, I might have lain down every now and then and looked along its surface; in which case I should, doubtless, have succeeded better.

The mention of Blackgang leads me to speak of the Isle of Wight, as the choicest storehouse for this kind of fossils in Britain.

If it be true, as some geologists aver, that the pearly "Vectis" was at one time joined on to our Hampshire coast, from which it afterwards broke off like a loose morsel from the side of a cake, that would go far to explain the abundance and beauty of its organisms. For, upon such a separation from the mainland, the industrious sea would have great opportunities of denuding the face of the cliffs and sucking the orifices of the chines. Also, this circumambient sea is peculiar in its character hereabouts; and the crystallizing waves impart an unusual lustre to rolling agates and jaspers.

The beaches of the Wight are circumscribed in extent, which is a great relief after the interminable ranges on the Sussex coast. The difference is as great as that between a quiet forenoon spent in the museum in Jermyn Street, and the laborious distractions experienced if we pass a similar interval of time amidst the endless rooms and innumerable cases on the first floor of the British Museum.

The village of Sandown, from the flag-staff on as far as Red Cliff and Culver, exhibits a tempting and instructive shore. In winter, it is better to commence at the dyke, and end with Red Cliff; but in summer, you may walk the whole range with advantage. Here, a couple of hours may be taken to as many miles; and if you do not shut your eyes, you must fill your pockets. But look with resolution for good ones.

From Sandown to Shanklin, in the opposite direction, it is scarcely inferior. It was once even better; but, lying so near at hand, it has been unceasingly hunted over; and to look for a fine pebble there now, is almost like hoping to find a hare sitting in your kitchen-garden. Many tons' weight of beautiful fossils have been gathered within the compass of a morning's walk here, in the last few years.

At the foot of Shanklin Chine occurs a singular collection of limestone pebbles, siliceous within, which are well worth examining. Extraordinarily-tinted choan-ites are here sometimes detected in the heart of most unattractive and mis-shapen blocks. I once asked a successful beaching lapidary there, how he *knew* them? He declared that he did not know them at all; but that whenever he saw fresh bonzes thrown up, especially if with a deep-sea mark upon them, he would break them all, and generally obtained a superb specimen among them. The objection to this murderous process was, as he confessed, that some of the finest were wont to be destroyed.

Latterly, I thought I discovered a way to recognize the presence of any such choanite. It is to look closely for a peculiar depression in the stone which is con-nected with the principal *air-hole* belonging to the fossil. But this method demands time and care.

Onwards to Luccombe, the beach disappears, owing, I imagine, to immovable depth of sand. At Luccombe Chine, it partially revives: but do not linger here. Go on to Bonchurch, and as you approach that lovely hillside, rouse all your energies. Here comes in a mile of sand, small shingle, and half-buried pebbles, which will put your skill to the test, if skill you have. Look out for symptoms of the "tubular" structure, and diverge now and then to the pools among those weedy rocks. Many a beautiful "actinia" in flint lies there snug enough, wedged in a dark cranny, and coated over with glutinous moss. You must really not be idle now. Do not even light a cigar, but hammer and delve and scrape away; only do not give in until you lay hold of something worth the while. Two zoophytes, of the agatine-siliceous kind, were picked up on this spot about a twelvemonth since, for which large sums of money were offered in Ventnor the same day. For one of this sort, an American gentleman, not long ago, gave eight pounds sterling. But there is in Brighton, or was till very recently, a pebble from the Rottingdean beach, for which fifty guineas have been offered to its possessor, and refused. I have not seen it myself, but I am told that it is a spotted pyriform choanite; the material, very beautiful agate.

At Ventnor, red gravel and diminutive carnelians, some of them not bigger than pins' heads, await the pebble-seeker. Occasionally an agate is found here which is "chatoyant," like an opal, but this is scarce.

I once tried the Freshwater strand, "ultima thule," but I only met with doubtful, ragged flints. The north side of the island presents on its shores a surface chiefly of mud and estuary drift, delightful to solans and sea-gulls, but not profitable for the cabinet of the mineralogist.

I ought here, in common gratitude, to say something of the beauties of the Garden Isle. But, though some of the happiest hours of my life have been passed there, I grudge to dilate upon the theme. I could not, if I would, convey to a stranger a sense of the unutterable attachment which I feel for its swelling downs and curling seaward bays. Much of the charm of our old English scenery, which is fast disappearing elsewhere before the railway and the canal, is still to be met with in this sequestered island. Here are uncultivated rocky slopes, proffering rare wild-flowers; ivy-mantled cliffs and moss-grown brambles; bays which in winter witness the howling storm, and sometimes the fearful wreck, but in summer are invested with the calm beauty of a moonlit lake. The inhabitants also, many of them "aborigines," have a primitive, unsophisticated character in the inland villages; while, on the coast, this is chequered by that roving disposition and semi-superstitious bent which are always to be found among sailors.

Oddly enough, the story of a pebble illustrates this trait of character. Some years ago, I was calling on a lapidary in Sandown who had done a good deal work for me, and done it well. His account was settled, and I was wishing him good morning, when he drew my attention to a remarkably beautiful pebble, ready cut and polished, and begged my acceptance of it. Now the stone, displayed on his board along with other specimens, could not fail to sell; and was worth *to him*, I should say, perhaps a sovereign. I was very unwilling to take it, but there was no refusing without offending an old friend; so it went home in my pocket. But somehow curiosity was stirred in my mind when reflecting on the circumstance afterwards. Why should he wish to make me a present? Or, allowing for goodwill taking so straightforward a course, why part with a stone of so much moneyed value? A few days after I dropped in again. My friend was out: but I found his wife, and stayed for a chat. Presently I referred, quite cursorily, to the gift which her husband had made me, "and," said I, "I wish only it had been something of less value." The gentler sex are sometimes more communicative than their mates. "Ah! sir," replied Mrs. —, "there was reason for giving you *that*, may be. Anyways, I'm real glad that it's gone from *us*." "Indeed! how is that?" "Why, that stone was an unlucky stone for us, and I wished it to go, somehow." Then I learned from her the following singular tale.

Two brothers, sea-faring men, were great friends of the lapidary and his wife. One day, one of these brothers found this pebble on the beach, and had it cut. It turned out, as I have said, a remarkably fine specimen. It was what they call a "deep-sea" pebble, and the "choanite" of a reddish hue lay exactly central in the bonze of white limestone. The sailors were both of them quite fond of it: and when they went to sea again, the handsome pebble went with them, and was laid "o' nights" on a shelf by the hammock.

Their boat was upset: one brother was drowned: the other righted the craft and got to shore, not without difficulty. Several articles were missing, having gone to the bottom; but the pebble was safe.

The survivor brought it back to the lapidary, and made him accept of it, being unwilling to retain it himself. The lapidary took it, and put it aside: but his better half found it out, and insisted that he should neither sell it nor keep it. So, he *gave it away*. This stone I afterwards parted with to a London collector: not wishing to preserve so sad a reminiscence. I took a fossil in exchange, which I almost immediately lost. So, nothing connected with the mysterious choanite was "lucky."

Perhaps every feature in "Vectis," from the matted festoons on its honey-combed rocks to the "witchery of its soft blue sky," predisposes a frequent visitor to approve also of its pebbles. But it is an unprejudiced fact, that in texture and colour they excel those which are found on the Sussex beaches. And while I think highly of our Devonshire jaspers, I remember a jasper-agate, gathered some years ago at Sandown, which might challenge all such stones of native growth in our London museums, and bear away the palm.

Many a calm summer's evening, many a stormy winter's forenoon, have I whiled away in absolute contentment on that most interesting shore. In the hot noons of July the bathing is delicious; and after your bath, the shadow cast by overhanging cliffs, and a fanning breeze from the sea, make a ramble onwards irresistible. But I found winter as pleasant in its way. When the gust roared, and chimneys smoked, and the heavens looked dark and disturbed, it was inspiriting to go forth roughly clad, and to watch the fitful changes where the weather-gleam strikes upon Culver; or to mark how that treacherous mist would come creeping round Dunnose, till the "white horses" rose on the sea, and the heavy squall which had long been packing in the offing burst with a wild cry, driving in a yet heavier element with tatters of oar-weed and splinters of broken dyke against the face of the immovable cliff.

After passing Shanklin, the style of scenery as far as Luccombe and the Fishers' huts, is quaint and picturesque. Here the "Gault" makes its appearance, and the hill-paths are in consequence slippery and dangerous. After this, succeed several miles of fairy-like beauty. And then, the stern grandeur of Black-gang, the coast-line trending away in a range as savage as that of Calabria. And last, the chosen home of the poet, sublime Freshwater, and rainbow-tinted Alum-Bay.

Having said so much on the pleasure of beach-rambling, it may be well to put tourists on their guard in two important particulars. The first of these touches personal safety. No one should set out upon what may prove a prolonged expedition without first ascertaining, *on Kalendar authority*, the hours of the tide for that day. There are sundry points on the coast of the Isle of Wight, for instance, where after a couple of hours' progress, on scientific thoughts intent, the pedestrian may find himself hemmed in by a jutting precipice in front, while the advancing tide behind has gradually cut off all hopes of a retreat on "terra firma:" and as the face of the cliff is seldom such as admits of being scaled, it will then fare badly with him if he be not "the king of fishes."

The other matter respects fatigue. Hearty exercise is desirable; not so that which results in utter exhaustion. My own experience is as follows:—I have climbed to the summit of Sca' Fell, and then walked back, footsore and fasting, to Keswick. I have swum a mile before breakfast. I have ascended Mount Etna at midnight in December's snow. But for sheer fatigue, pulling at the muscles, and drying up the marrow, there is nothing equal to ten miles of stiff shingle, while a foggy north-easter currycombs your face, and perhaps your shoe-leather has been laid open by the edge of an inauspicious flint.

If in addition to all this, forgetting "Cording," you have sallied forth *non bene relictâ parmulâ*, your condition is by no means enviable. Therefore, whoever intends to go beaching in earnest should look to his outer man, and carry with him certain creature-comforts, not omitting to include in these a supply of good tobacco.

Yet one word more upon this head. Have a due regard to the well-being of your eyes. They are, as you will not fail to acknowledge, the working party on these occasions. Manage how you will, they *must* be subjected to a great stress and strain. Give them half an hour's respite, if possible, while you sit down and inhale the "weed." It is unpleasant to find, after too intense a service on the part of your "daylights," as you return homewards along the meadow-path, that a beach you have left two or three miles behind you is all *coming back*, whether you like it or no, upon your organs of vision.

On the whole, summer is the most fruitful season for stocking a cabinet, on account of the more powerful illumination cast upon the objects of your search. But in winter the beach itself is very likely to be in better condition, for it is oftener agitated, and many specimens will come to the surface which in calmer weather lie buried under sand. It is important to train yourself to attentive observation, and after long habit the ear will listen intuitively.

If you hear the sea make a dull, booming noise during the night, be on the look-out two or three tides later, when the first shingle now thrown up has sifted and settled down. Large pebbles will then be found on the top of all, the process being much like that which takes place on shaking a basin of lump sugar.

But it will be well for you to be always on the alert. The seaside elements must be coaxed and humoured, if you hope to get anything by their agency. A beach itself is exceedingly capricious. On some days you may walk for miles and remark nothing worth picking up; perhaps the next day, over the same ground, so great will be the profusion of fossils as to suggest the idea that since your last visit a petrified shoal had returned to life, swum into the bay, and there been once more stifled in gurgling lime, or liquid silex, and penetrated by the metallic "moss."

"The sea yields its harvests without sowing or planting," and seaside scenery is exquisite if you have come upon it where it is unsophisticated. Happily it is in such sequestered "Edens of the *western* wave" that the most desirable pebbles are met with. I have gathered rare specimens where the dor-hawk was wheeling overhead, and saucy sea-gulls screeching almost in my face; while a cunning crow would watch from a projecting ledge of rock to see whether I was about to capture, and afterwards throw away, anything which might serve him for a luncheon.

The terraces of Margate and Ramsgate are invaluable to the tired artizans of London seeking their well-earned recreation; but no poet could venture to affirm of them what Scott said of "Brignall banks," that they are "wild and fair." But quit these populous thoroughfares, and get away to pearly Beachey Head, or roam the lone strands of Yorkshire or Devon, or go and lose yourself among shadowy nooks and gleaming bays in the sweetest of all islands, and you will possess the genuine colour and scent, and music and mystery of the sea, as the Creator has framed and blended that wondrous element.

You need not look for pebbles unless you like; sometimes it were better not. But saunter down to the water's crinkled edge, and inhale that indescribable odour from old rock, slippery now with dulse and ribbon-weed—Piesse and Lubin distil nothing to equal it—and con the page in Nature's volume which lies open before you; it will never give you a head-ache, nor a heart-ache either.

Here, after presently stripping to the elbow, you may surprise a hermit-crab, or catch the spotted "goby" in his dark saline pool, but please let him go again; you ought to give him his freedom, if only out of gratitude; for all this time your own tired body is being braced and refreshed, and your mind, yesterday jaded and careworn, is winning back its elasticity under a sense of blessed repose.

I will now mention some peculiarities in the nature of our beaches, which I have noted from time to time in the book of my experience.

A BEACH is supposed to be a permanent accumulation of pebbles and small shingle, due to such cliffs or banks as are near at hand. And this general definition would be so far scientifically correct, if nature were stagnant like the surface of a deserted mill-pond. But, as Nature is always at work, always shifting the scenes, the above definition, although it sounds like a truism, does not hold good in either of its clauses. The accumulation which we look upon is *not* permanent, nor is it always due to a neighbouring source. Sometimes the collection of stones which we find heaped up in an immemorial bay has been washed round—much of it, perchance, recently—from a distant quarter. And this might be taking place under our very eyes, and we not know it, just as we do not *see* the grass grow. Except in very calm weather, beaches are continually travelling, and they travel rapidly along certain lines of coast, until they arrive at a terminal point or headland. To get round this latter takes time, but they do get round it at last, with the help of strong gales prevailing in one direction, and under the rushing tides of an equinox. Now if we suppose a few acres of shingle to be once fairly carried past such a promontory, and discharged into the curve of an elliptical bay, they will be locked in, and may, perhaps, never get out again.

I have myself no doubt that Sandown Beach is fed and replenished in this manner: and the bay being deeply indented with the points of Culver and Dunnose, prominent at opposite extremities, egress is difficult. Yet most persons might naturally suppose that whatever they find on the beach here is due to its overhanging cliffs.

Some of the best pebbles travel fast, and are, in fact, migratory, till they reach such a "nidus" as I have described. This is owing to the rounded form which fossils usually assume. Angular lumps rather remain stationary. Jasper I suppose makes few excursions: while the agates and oval choanites are continually on the move, and may have visited a dozen different localities before they are picked up. I have hunted a bit of yellow agate in this way for above a mile in the course of a few weeks, dropping it again on purpose, and never finding it in the same place.

Beside the above real motion of a beach, there is an apparent total change for a time which is by no means uncommon in exposed situations.

This is, when an inshore gale drifts the sand and conceals for a day or two the entire mass of shingle. It is a provoking circumstance when it occurs, but it seldom lasts long in its effects. The sudden re-appearance of half a mile of solid beach, as if by magic, on a change of the wind, has sometimes surprised me in spite of past experience.

In average weather, when a beach is not migrating, the rise and fall of the pebbles in the same lines is very observable. The only change then induced in a stratum of shingle is by the suck and draught of the sea. This latter, however, is unceasing, as may readily be tested by bathers; everything which you set your foot upon under the wave being in a gentle rolling motion. It yields an amusing diversion in the water, to play the pebbles about under one's feet. The effect of it, however, is not so quaint as that which I remember experiencing many years ago in Loch Long, where are no pebbles, but plenty of mud. Here, wading out in the shallows to arrive at swimming depth, I put one foot after the other on the backs of some unsuspecting flounders lying on the sand, and laughed aloud at the tickling sensation as they glided away from under me.

But all *motion* is a source of interest and charm, especially while we are on the sunny side of forty. I believe that this forms one feature of delight in the capture of winged insects. Exquisite as is a good British collection, the dry side of the science, whatever learned entomologists may say, is not the most inviting, and would attract comparatively few persons after a time.

In like manner, who would care for cover-shooting, if all were sitting-shots? But the "runners" and "fliers" impart a zest.

Even here, my friends the seaside pebbles have a claim to please: for, although inanimate, they are not altogether devoid of motion, as I think I have shown.

CHAPTER V.

Concerning Certain Petrified Zoophytes, Now Generally Supposed
To Be Extinct.

The petrifaction of an animal form or structure has always appeared to me a most singular process in Nature. The change of such substances as cellular tissue, or fish-bone, or cartilage, or a light, leathery skin, into minerals such as silex and agate, seems at first like magic. Scott, when describing the "foliaged tracery" in the east oriel of Melrose Abbey, tells his reader:

> "Thou wouldst have thought some fairy's hand
> 'Twixt poplars straight the osier wand
> In many a freakish knot had twined,
> Then framed a spell when the work was done,
> And changed the willow wreaths to stone."

But there is no magic in Nature; and when we meet with undoubted phenomena, however strange they may appear, in any department of her realm, we must behave ourselves like matter-of-fact persons and set about accounting for them logically. In a genuine petrifaction there are two ways of doing this, or, in other words, there are two different modes in which Nature may have acted so as to produce the result which we are considering. One of these is that of a simple *substitution*; the particles of one substance being gradually removed, and those of another taking their place. This must have been the process which obtained in many well-known instances: as, for example, when the thin, leathery shell or husk of an echinus is found perfectly rendered, both as to its general outline and minute markings, in solid flint or limestone. We feel quite sure that the original husk has perished; but here is its second self.

The other mode is that of actual *transmutation*, wherein no part of the original substance is destroyed or removed, but its physical conditions undergo a total change, by means of either infiltration, or crystallization, or perhaps both of them combined. This process has obtained in such cases as a petrified shark's tooth, or a fossil trilobite from the Wenlock.

The tooth, which was once bone, is now a kind of metallic stone: the trilobite once, as we suppose, shell and cartilage, is now a something between limestone and cast-iron. And this change is very wonderful; but, as was said above, there is no magic in it, no more than in the hardening of an infant's skull, or the solidifying of the arm and leg-bones by the absorption of phosphate of lime. Besides, the stone and metal now present are still something different from what we ordinarily mean by those terms: the petrified tooth is not like a flint-stone, neither could you cut horse-shoe nails out of the trilobite.

In fact, every *real* process has a character of its own: a something which distinguishes it from any other, despite of general resemblances up to a certain point.

Petrifaction, then, being assumed, as an operation which has really taken place after one way or another, it falls next to be considered, what was the original nature of these fossils? I mean the objects which we find thus preserved in our sea-side pebbles. What was the department in Creation to which they belonged? Was it *animal* or *vegetable* before it thus became altogether *mineral*?

For myself, I will at once say that I have no doubt it was animal; but not an animal of a very high order. The specimens which I now possess, and which are all of them chosen, some of them *rare*, I should assign to such creatures as the "zoophytes" or "polyps," both radiated and globular. And I know of nothing, among many hundreds of specimens gathered from a dozen different beaches, which presents the evidence of having belonged to vegetable organization, with the exception of sundry varieties of petrified wood, which speak for themselves, and could not be mistaken by anybody.

I will now state two or three reasons, which to my own mind are conclusive, for the above decision.

In the first place, then, the structures here perpetuated in stone are of great delicacy, and they have been immersed in ancient seas, as is testified by the localities in which they are obtained. Now vegetable structure as fine as this, if immersed long enough for any such change to come in question, must have utterly perished by maceration, and then the petrifaction could not have taken place. I know that part of the stem (and I think fruit) of one species of "conifer" has been found in the Isle of Wight in the condition of a fossil; but this belonged to a hardy class of plants, and the lobes or plates which composed its bark and husk are themselves highly siliceous, to say nothing of the presence of iron in the rind of most of these stems. So that the process would be a long one, and the fibrous material of the tree would stand it well. But in these pebbles some of the threads or tubes run from the size of small twine to that of the rays in a spider's web; and no vegetable substance with which I am acquainted, excepting the filaments of "asbestos" (which is a vegetating *mineral*) in rock-crystal, could abide and retain its form, so as to allow of the changes by infiltration or otherwise which have passed upon the original structure.

If it be said in reply to this, that we have the exquisitely delicate "dendritic" markings, as of leaves and filaments of shrubs or sea-weed, in the heart of the white chalcedony "mocha-stones" from the East, the answer is evident: these are not *really* vegetable traces, but only *resemble* such in their configuration and colours. They are simply shoots and ramifications of a metal, — as iron or manganese. Those in the "weed-agate" of India, which exactly resemble sprays of fine sea-weed, are produced by "delessite."

Indeed, it is both diverting and instructive to observe how Nature permits, and even seems to abound in, curious coincidences and striking resemblances between things of entirely diverse character. The dried polyp, called "encrinoid echinoderm," bears a wonderful likeness to one species of Indian corn. (See the plate at p. 137 of Mr. Rymer Jones's beautiful work on "The Animal Kingdom.") And the other day, when I was enjoying a leisure hour in the British Museum, I suddenly remarked that the "carapace" (back-shell) of the splendid fossil specimen of the "Holoptychius nobilissimus" in one of the cases might serve for a sketch of the back of a *capercailzie*, where the grey and purple feathers overlap one another. Yet here is no *real* connection whatever. Only Dame Nature had gone to play.

Secondly. The preservation of these "polyp" forms, in the manner in which they *have* been preserved, seems to me to be due to a feature or circumstance which is strictly animal and not vegetable. I refer to the fact of the creature, while it was alive, *inhabiting a house*, a house built by himself, or emanating from his own substance. For, just as we could know but little of the existence or habits of "shellfish," were it not for their *shells*, so I think we may assume that the *choanite* must, when alive, have dwelt in a tough, horny "coperculum," answering to the shape of his body and the number of his "polyps" (if he was a *compound* creature), because otherwise he would have been like a jelly-fish or naked slug, and his "polypary" could not have been preserved in a stony fossil. Of course, whatever was merely flesh, or adipose matter, has long since perished; but the *house* or shell in which it whilome dwelt remains. Thus, the "echinus" built himself a *dome*, such a residence as a *hedgehog* would require to live comfortably in, and through the various orifices of which his spines could be protruded at pleasure. The "ammonite," being shaped like a *snake*, preferred living in a shell of that form, where the creature when coiled up was safe. The "alcyonite" had a more exquisite taste in house-building, answering, we may be sure, to the complex and beautiful structure with which the great Creator had endowed him. His home was a palace, containing long galleries and secret doors and wheel-windows; and here some of the delicate tubes are fringed at their extremities like the petals of a flower.

The antiquity of these formations may be very great, we can scarcely tell how far removed from our own era. For, while the zoophyte itself is of a perishable nature, we are acquainted with no substance more durable (if we except the gems) than that calcareous matter of which these tubes and plates were formed, when once it has been subjected to processes of infiltration by crystallizing mineral and metallic oxides.

Now, I think this argument a very strong one; in fact, although simple, almost unanswerable. For no *plant* dwells thus in a house. We have the plant itself, but nothing more; and if this be not capable, and I hold it to be incapable, of sustaining the most vehement mineralizing process in the crucible of Nature, its history must be a brief one, and excepting in the dark "lithographs" of the coal-measures, its memory must pass away. I have already allowed an exceptionable case in favour of the *conifers*, which, be it observed, nowise resemble anything portrayed in siliceous pebbles, and it is remarkable how much this class of plants predominates in the COAL.

I have always been suspicious of what are called "vegetable petrifactions." I examined those at Tivoli, near Rome, in the year 1845, and I made up my mind that they are simple *incrustations*. In like manner, many of the buildings at Pæstum are constructed with a kind of "travertine" taken from the bed of a neighbouring river, and which rapidly incrusts any solid objects submitted to the action of its waters. But the truth is, the vegetable pipe or "straw" remains for a while, owing to the silex which entered into its composition while the plant was growing. After some years the *straw* decays, and there is a hole or depression in that part of the pillar or pediment. On the other hand, a calcareous "menstruum," imbibing silex and iron, hardens into a substance which, like the best mortar or cement used in building, will sometimes outlast even the blocks of limestone or oolite which it was put to bind together.

Thirdly. Animal organization, such even as these polyps possessed, renders the phenomena much more intelligible. Our best authorities in such matters tell us that "*insects* have neither lungs nor branchiæ; but in them the air passes into *a system of tubes*, whose structure resembles that of an elastic webbing." And again, "The *annelids* possess an *uninterrupted circulation*." And again, "In the 'Nymphon' and 'Pycnogonum' molluscs, which are crustaceans having considerable resemblance to certain of our field spiders, the intestine *penetrates to the very extremities* of the feet and claws."—*Animal Kingdom.* Now here are cited some of the very desiderata which I should have named, had I been asked what conditions were needed *à priori* for such petrifactions to occur. I will only add, under this head, that a fine *annelid* occurs in the blue agate off Eastbourne; that a "myriapod," which is among the chromo plates of this volume, has all the characteristics of *insect* life and motion; and that a *spider* is the nearest thing I know of, in some respects, to what the "choanite" must have been when that mollusc condensed himself from a cylinder to a sphere. Perhaps, however, the strongest clause in this part of the argument may be drawn from the "sponges." Here the creature itself, wonderful to relate, is a *viscous fluid*, and the intricate mansion which he inhabits is a globose, horny skeleton, perforated with endless small tubes opening into wider galleries. There was, however, in the perfect animal, I am assured, one main central cavity, which gave strength and unity to the entire fabric by the plan of its walls, and, perhaps, by a main valve. Throughout the whole of this hydraulic system the sea-water, on the circulation of which the zoophyte depended for life and health, could be pumped to and fro at pleasure. And, evidently, when the "habitat" of such a creature was suddenly invaded by a siliceous crystalline solution, extinction of the animal and a petrifying investment of his abode would be simultaneous.

Lastly. If the objects here petrified had been vegetable in their extraction, should we not, with the aid of the microscope, be able to identify them? But this I have never yet succeeded in doing; yet all the petrified "woods" are well known. I have myself obtained slabs of the "acacia" from the coast of South Devon; of the "beech," in Sussex; of coniferous wood almost everywhere. And, what is more to the purpose, though the petrifaction in such cases is deep and perfect, no one looking upon it could doubt for a moment that the original structure had been that of wood from a forest-tree. Agatized as it is, and penetrated here and there by metallic colours, and shot with rays of jasper, the lines in its fabric reveal *the texture of wood.*

I may mention here, that every one who walks our beaches, with a view to the collection of fossil specimens, will do well to carry in his or her pocket a good lens, of large external diameter; mine measures about two inches across, and I may truly say it has saved me a world of trouble, besides affording me much satisfaction at odd moments in the scrutiny of pebbles of different kinds and textures.

Before closing this chapter I may be permitted to draw the reader's attention to a theory held by the late Dr. Mantell. I cannot, at this moment, lay my hand upon the volume in which it occurs, but I am pretty sure it will be found in his "Geology of the Isle of Wight:" a book which, for elegance of composition, and sound information, can hardly be too much commended; though a resident lapidary in Sandown *did* once say to me, while thumbing the pages of a well-worn copy, "Ah, sir! if the Doctor had come *here* and stayed a week instead of listening only to what those fellows told him in Ryde, I could have shown him something which he doesn't seem to know, as to how the bit of coast runs hereabouts." Dr. Mantell's idea was this: he held that when a mollusc was subjected to the first stage in the petrifying process, there was, in the dying of the creature, some effusion of blood (or *quasi*-blood), and that this, being the very pith and strength of the animal's system, would, in many cases, tinge the future stone indelibly. He carried this notion so far as to assign some dark blotches, apparent in the masonry of a wall, to such a source as being their most probable cause; and he gave to the thing itself the graphic title of *Molluskite*. In this view I will only add that I am inclined to agree with him; and in my "myriapod," already referred to, there is a blood-red spot which pierces through the stone, appearing on both sides, and which I at first supposed to be a piece of "shell-lac," but I now rather regard it as the trace left of himself by some marauding "pholas," who, after drilling a hole through the solid pebble, found his own grave there.

CHAPTER VI.

On The Litoral Labours Of The Ocean, And On Seaside Sports.

The more I consider the various phenomena occurring from time to time on the dry land, and the more opportunity I have of observing for myself simple facts in geology, the more I am struck at once with the truthfulness and the *unexpectedness*—if I may use such a word—of the assertion which Moses makes in the 1st chapter of Genesis, that the appearance of the "dry land" was due to "the gathering together of the waters unto one place."

The assertion has an unexpected (*à priori*) character; for continents and tracts of dry land do not, on the face of them, suggest the idea of any recent presence of incumbent masses of salt water—perhaps miles in depth. But it is eminently truthful, for it is a key to many an abstruse problem in Nature, and a confirmation of every sound and enlarged view of the past history of this globe.

The ocean has been busily at work—in old times, inland; in later times, coastwise; in all times subterraneously. This last point is proved by the volcanoes, and that in a twofold argument. Such volcanoes as are now extinct, are so because they have lost all communication with the sea; such of them as are active, are so because they draw supplies of salt-water from the nearest part of the ocean, and this they can only do subterraneously.

But in speaking of the labours of the ocean, I shall confine myself to the seashore, as the scope of this little volume does not go beyond that region. The point where sea and land meet is the critical point for all observers of Nature. Here the disciple of geology should serve his apprenticeship, and if he cannot accumulate facts, and glean a kind of inspiration here, he cannot do so anywhere. Moreover here, better, we think, than in any inland scenery, Man can muse and meditate. That ever-varying curved line of moisture on the shore depicts the fluctuating changes which momentarily visit his "little day;" the tide running in is the flood of his early life; the tide running out is the ebb of his declining years; the vast sweep of the coast, backed by the upland ranges and everlasting hills, and itself only lost to sight in the far horizon, tells of a steadfast future, an immutable eternity.

Above all, those who desire to note epochs in the flight of Time, and to set up way-marks in the Earth's chronology, must study the line of the sea-coast, the ancient and the modern, for here, if anywhere, the dial-plate is uncovered, and the shadow of the gnomon may be traced through some seconds of the enormous day which has witnessed the existence of the heavens and the earth.

I have already, in my opening chapter, remarked how the sea brings down, in the course of ages, many a pebbly beach from cliff and causeway. But I am far from assuming, therefore, that *all* the pebbles of a beach come from the land. The usual bottom of the sea is, indeed, no pebbly shore; but there are many submerged rocks of sandstone and oolite, out of whose ribs and crevices, from time to time, fossils may be washed, just as our own chalk-cliffs, the main resort of the siliceous pebbles, were themselves laid down in deep seas. And the salt water, which is always acting gradually to dissolve certain rocks, when it removes portions of these from the edges and promontories, will occasionally bring their contents to shore. But these results, although interesting to those who maybe searching after pebbles (and a deep-sea pebble is a prize), were not what I pointed at when I spoke of the labours of the ocean.

I have frequently walked the shore, and observed the colour of the waves, after what is termed a "ground swell," which had lasted, perhaps, for thirty or forty hours. The cerulean hue is then gone, and to it has succeeded, in certain localities, an opaque chalky tinge, showing that the water is now heavily charged with lime. Also fragments of shells rolled together are united with heavy masses of sand, and sometimes of broken pumice-stone, and a kind of rough marl is rapidly formed, and left on the beach. After a gale, and succeeding "swell," I have met with these imperfect boulders, varying in size from that of a man's fist to some larger than his head. At the same time, any low ranges of littoral rocks become crusted over with the superabundant lime, being more than the waves will long hold in solution; and a coating is thus given to such rocks which is sometimes as hard as is the native limestone itself, a few weeks' exposure to the sun and air sufficing to effect this.

Some fourteen years ago, I had an opportunity, when in Sicily, of examining a portion of the coast between Messina and Catania, and I regret that I did not avail myself of it more heartily. But I have seen M. Quatrefage's book on this subject, and his observations, most carefully and laboriously conducted, may almost be said to close that part of the subject, as far as any prospect of eliciting fresh information is concerned. I think he measured some of the long reefs, and the evident increase *by incrustation* extended for many miles of the coast-line, and was of considerable thickness. I have observed the same "masonic" phenomenon off the coast of St. Andrews, in Fifeshire, when swimming out among the weedy rocks, and afterwards climbing to the shore. I then thought it was the work of marine insects, as I had heard of molluscs building causeways of tubes of limestone, but I incline now to think it was the sea doing it, as they say, at *first-hand*. The sea, however, does a great deal in the same line at second-hand, by means, chiefly, of two species of zoanthoid polyps. Of these little creatures, one kind constructs the bases of coral islands, and another the summits (at the least) of the madrepore reefs. It is impossible to doubt that by labours so patiently carried on, and so widely diffused, some beneficent purpose is aimed at and attained by Providence. Perhaps they operate, finally, to warn the shipping of adventurous merchants from entering certain dangerous bays and straits; and if the humble madrepore has got a bad name through this, as though he had made the danger, he certainly suffers unjustly, for what real difference, as to ultimate and assured safety, can a few inches more of water in such places make? Far better to raise an impassable bar across the way at once, and proclaim, "no thoroughfare!"

Then the ocean supplies a great market, much the greatest in the world. I do not know the proportion of persons in Central Europe who live on fish to those who live on meat, but I think, in both Northern and Southern Europe, the former exceed the latter. In Connemara, on the western coast of Ireland, and in the Scotch highlands of Argyle and Inverness, fish is decidedly the staple article of diet, as far as animal food is concerned. I believe the same is true of Cornwall and part of Devon. The herring, the pilchard, the haddock, the cod, and the salmon—to say nothing of the sole and other flat-fish—feed millions of persons, and, with the help of oatmeal, barley-cakes, and (in a good year) potatoes, feed them well. Shell-fish are also no mean item, and are sometimes the most refreshing of all dishes, especially at the supper-table. Now the sea, it must be remembered, does all this for man without harrowing, ploughing, or sowing. The fisherman's net may be said to *reap*, that is all.

Again, if jewels be of any real value, what is the value (among such) of a collar of faultless pearls? what is there among minerals, so pure, so exquisitely beautiful, whether we regard their tint or their form? These are the spontaneous production of a humble shell-fish; some say, an offering from the creature when he has been wounded. If so, men may here learn a lesson in kind, and return good for evil to those who persecute them.

Again, the ocean is *our* defence: long may it prove so!

> "Britannia needs no bulwark,
> No towers along the steep,
> Her march is o'er the mountain wave,
> Her home is on the deep."

No doubt, since man is a creature who *lives* upon the dry land, the land is on the whole the most valuable; but what would the land be without the sea? Nay, to go no further, how could we exist in summer and autumn, if deprived of the sea breezes? For myself, though naturally fond of field-sports, and delighting in botanical and entomological pursuits, I know of no treat equal to that of a seaside ramble in the month, say, of September or October.

Now let us consider how very little mischief the ocean does us. Much fewer people die of what are called "casualties" by sea, than of those by land. The truth is, we always hear of such as are lost at sea, because of the loss of *the ship*, which involves a question of insurance-money; but of deaths inland, especially in far countries, and if belonging to strange peoples or savage tribes, we perchance scarce hear at all. YEH, the Chinese Commissioner, killed, it was said, 70,000 persons in the course of his brief career, for *political* reasons. How long would it take for 70,000 persons to perish by storms or accidents at sea, in the usual course of things, in one small part of the globe?

A wreck, it must be allowed, is a terrible thing; but so is a house on fire, or a flood up the country.

The sea, moreover, labours to help the land in other ways, some of them singular enough. Many thousands of poor unproductive acres have, in these last ten years, been rendered rich and fertile by supplies of GUANO. And whence did we get the guano? why, from the *sea-fowl*, and from a barren rock in the bosom of the seas! This is no trifling benefit to reap from the "desert sea."

What a beautiful thing is *glass*, and how indispensable it has become to us! Perhaps no other material can be insured to be *perfectly* clean and pure for drinking out of. For many years our glass-works depended mainly upon a constant supply of *kelp* from the Orkneys, and even now they cannot dispense with the sea-sand.

The porcelain mills in Staffordshire and elsewhere are very glad to obtain a cartload of pebbles from the beach. These go to the "crushing" department; and if among them there are, as there are pretty sure to be, a dozen lumps of chalcedony, the material of the next batch of teacups will be unusually fine, provided some one who understands them *picks* the stones first. Indeed, if I had no preferable occupation in this world, I have often thought I would collect the rough agates and chalcedonies from sundry localities I wot of, and fabricate a *biscuit* of my own, as the king of Saxony does at Meissen. But though I have not the leisure for this, others may have; and so I mention it here.

Especially I incline to think that a splendid kind of "Wedgewood" ware would result from the crushing of certain jaspers, for I suppose their colours would not fade in the furnace.

Then, the seaside visit, I must not omit to mention, can be enlivened by sundry local pursuits and amusements. Besides the pleasures of a sailing-boat, and a run with the "dredge" and the "dipping-net," there is the exciting march of the Shrimper, knee-deep in the wave, pushing the hoop-net before him, and every now and then halting to fill his front pocket with the silvery jumpers. If there are rocks near at hand, there will be *lobster-pots* to visit; and the habits and deportment of a live lobster are among the most curious in creation. In agility and cunning he surpasses even a salmon. Also, for the benefit of those who at all regard what they eat—and he who disregards it is a goose—I may just venture to hint that a *real* lobster-salad (London confectioners have a way of selling *sham* ones,) is a dish worthy their serious attention: Barclay's stout being not a bad accompaniment.

But the above will sound to some persons too "Epicurean." O gentle reader, do you love moonlight? and if you have ever admired the reflection of that planet in a lake or river, what will you say to it when you contemplate it in Sandown Bay? Culver on one side, looking as if it were of green glass, and the cliffs of Shanklin on the other resembling walls and pillars of porphyry! Or is your taste for the sterner beauty of storms and angry seas? Then visit Blackgang Chine late in winter, and you may "sup full of horrors." The appearance of the waves below, as they come in over that fatal "race," and the aspect of the earth and the heavens above, when the lightning darts from "St. Catherine's head" and sweeps like a destroying Angel down the chasm of the Chine, yield together perhaps the grandest picture of desolation and terror that English scenery ever shows.

There are persons still living who are unwilling *to speak* of the fearful tempest they witnessed on that coast when the Clarendon was lost.

Lastly, to return to our "pebbles," the sea is an indefatigable agent in the partly mechanical, partly chemical, work of infiltration; a process to which both the fine texture and varied colours of these agatized fossils are mainly due.

But of this I must treat in the chapter which follows.

Enough, for the present, of the ocean itself—of its labours and its sports. But, as some readers love *a comparison*, and hold that every theme grows dull without this, I will quote from the lips of a great traveller, whom it was once my luck to meet, his opinion of the rival claims of the Desert, to that admiration which we islanders lavish on the heaving Ocean and the winding Shore. I cannot pretend to remember his very words; but he was eloquent as the son of Laertes, and he made me long to visit the East. He said that the Desert was "another world," more marvellous than this of our land and sea: it was a home and a domain, like the former; yet was it waste and boundless as the latter. I asked about the *sands*. "Vast, beyond computation." "And the material?" "Powdered quartz, all of it. Quartz mountains, crushed, and pulverized, and sifted!" The reddish hue is from the peroxide of iron. No particle of anything like organized matter has ever been detected in its composition. The caravan-camels which drop and die daily on their hot march, as they have done for centuries, leave their skeletons, after the vulture's inquest is over, to bleach upon the surface; and, in the course of a season or two, these bones must turn to an impalpable powder: but that does not mix with the sand, or phosphate of lime would at once be found on analyzing it.

Then he spoke of the sunrise, and the glowing sunset; and of the delicious hours of night; and of tent-life in the Desert; and of wandering Arabs, who revere the grave and silent man; and of the charms of an encampment in some green "oasis;" and then, strongest commendation of all, he said, "I am going back, among the children of the Desert!"

CHAPTER VII.

Concerning Certain Natural Phenomena.—Origin Of The Diamond.—
Formation And Colouring Of Gems.—Infiltration Of Pebbles.—
Cause Of Transparency.—Integrity Of The Fossil-Nodules.—The-
ory Of The Shattered Flints.—Concluding Observations.

The charm which wins and rivets our attention in such pursuits as those of Geology or Mineralogy, is not that the phenomena which we meet with are capable of being classified, and of forming a scientific system. All this is, no doubt, one day instructive and interesting; but it was an afterthought, the result of experiment. The first charm lay in that silent mystery which broods over every part of creation, a veil as yet unpenetrated by Science: it lay in our instinctive consciousness that Nature, in what are, perhaps, her simplest movements, still transcends all the master-pieces of Art.

Take, for instance, the crystalline gems. These are the most remarkable substances with which we are acquainted; they are also among the most simple. Chemically speaking, the metals and gases may claim to be regarded as simpler bodies; but then it must be remembered that we do not meet with them in nature thus unmixed. Gases vary, both in volume and character, every instant; and ordinary metallic ores are penetrated and disguised by foreign matter; whereas faultless crystals, once perfected, appear to be unalterable.

Some of these, as the DIAMOND, would almost seem to be an elementary substance; and yet this can hardly be the actual case. The philosophical account of a crystal, as "some substance, all the particles of which, being free to move, have been operated upon in the way of a chemical, perhaps an electrical change," certainly makes against it; for, according to this definition, such "substance" is a first desideratum, without which we cannot have the "crystal." Thus, oxygen and hydrogen, blended together in certain definite proportions, yield the fluid substance, WATER; and the "crystal" of water is ICE. In like manner, common white carbonate of lime, and the fluor spars, whose ingredients we know, are in daily process of formation.

The philosophical idea, so accurately expressed in the above definition, is, doubtless, correct, if once we assume a past history for some substance, and take the crystal before us as its result. But, whether all bodies which we now meet with under the form of crystals *have, in point of fact, had such a history*, is quite another question.

The diamond is universally held to be pure CARBON crystallized. Its high refraction indicated this to Newton long ago; and the proof has since been given twofold: for the diamond scales off and evaporates under intense heat, and its dust has *carbonized* iron-filings, turning them into steel.

Further, this carbon is supposed to be of vegetable origin; and, if so, it would seem to follow that there must have been plants, and perhaps coal-strata, before there could be any diamonds.

But *what was the process* actually carried out in nature? No one has hitherto succeeded in obtaining this gem, answering to the conditions of a "brilliant" in water, lustre, and weight, by any attempted method, from the base of carbon. The French chemists have now, for some years past, been experimenting upon *boron*, by means of the voltaic battery; and they have, it seems, taken out of the crucible sundry small, pale crystals, which are found to be nearly as hard as adamantine spar (a variety of *corundum*); but they cannot show, as the fruit of their labours, a diamond weighing one carat, and worth, according to the tariff, eight pounds sterling in the market of Europe. Indeed, if this boron be a mineral intermediate between carbon and silex, their gems will assuredly partake of the character of rock-crystal, a stone which has no affinity with the diamond whatever.

The process of Nature, therefore, being unknown, if indeed it ever took place, the question fairly arises—whether diamond be a derived crystal, or itself an original type of created matter. Or it might be put thus: "If diamond be the purest form of carbon known, what is carbon then but diamond debased?"

Neither does this exhaust the argument. The ruby owns a matrix: the pearl grows in the mother-of-pearl. What is the nature of the diamond-rock? Is it a dark conglomerate? or is it diaphanous?

We know how diamonds are obtained: how they are picked out of the crevices of certain rocks, and washed out of the sands of certain rivers—in the Carnatic, and Brazil, and Borneo; but we do not seem to be much nearer to the history of their parentage.

Again, *what is* their "crust" or coating with which they are always found enveloped? Is it an integral part of the stone, or is it adventitious? In the best diamonds this crust is of a greenish hue.

Some say that while the gem itself has *positive* electricity, its crust shows *negative*. If this be so, the last question is answered—the crust in that case cannot be an integral part of the stone. The only absolute reason I know of for concluding the diamond to be a *derived* substance is, that it is laminar. The Eastern lapidaries, as it is well known, will sometimes *divide* a stone by striking it a sharp blow in the cleavage.

After all, the most extraordinary property of the diamond, as pure carbon, is its weight. It is fifty times the weight of refined charcoal. How was the element of carbon condensed and inspissated thus? Was it by the action of LIGHT? This is the vegetable analogy; as we see in the case of all growing plants: did it hold here?

If we were to inquire how oriental sapphires, including the ruby, the blue sapphire, the emerald, and the amethyst, are formed from CLAY, that clay which exists in the granite rocks, the difficulty would be nearly as great. We do not know at all.

But there is some satisfaction in having undoubted proof of the nature of their "base." And the obvious evidence upon this point is very simple, and prior to that of chemical analysis. My own attention was first drawn to it many years ago, in a casual remark made by a friend. We were handling some crystals of the *white* sapphire, a stone of little value. "These look very like glass," said I. "Yes, but you may always tell them from glass by their *coldness*. Touch one with your tongue." Then followed the inquiry. "And why is it so much colder than glass is?" "Because the 'base' of the sapphire is CLAY, and clay is a very cold substance."

Let us now pass on to consider the question of COLOUR. In the existing varieties of gems, we have all the colours of the *spectrum* perpetuated and vivid.

What is the source of these colours? And how are they blended with the solid crystal?

There seems to be no reason for doubting that the immediate source or cause of colour is the presence of some oxidized metal. All the colours which Nature has impressed appear, as far as we can trace them, to be due to such a presence. Opaline tints, and those of the "cat's eye," are an exception; being the perceptible result of a peculiar texture and configuration; so also are the iridescent hues on a *soap-bubble*, which are probably caused by polarized light. But it is *a metal* which makes the bark of certain trees and shrubs to glisten; aided, in the case of the birch and wheat-stubble, by particles of silex. It is a metal, absorbed from juices of the soil, which gives their tints to flowers, and their deep tinge to fruits. It is a metal which dyes the plumes of the king-fisher; and the gleaming scales of the dragon-fly and diamond-beetle owe their brilliancy to metallic lustre. Why should this all-pervading law vary in the crystals?

We know that metals proper affect crystalline forms, assuming them spontaneously; and we know that the gems have come in contact with metals. *Iron*, whose rust is of a reddish hue, enters into many of them; and in the *red* stones it is said to be abundant.

Probably, the finer permanent tints are due to *gold*, in infinitesimal proportions. Professor Faraday has said that the ruby-tinted glass called "Bohemian" derives its colour from gold in a pure form, finely attenuated, and not from any chemical combination of that substance.

A white diamond has far more lustre than one which is yellow or violet-tinted. But in all the sapphires, the deeper the hue, the more brilliant and valuable the stone. Hence the technical term for the colour of a diamond is its *water*. In all these gems the hues (whatever their origin) are homogeneously united with the crystal; and the *modus operandi* of Nature is a profound secret.

When we descend in the mineralogical scale, and come to such stones as agates and jaspers, the process which has been followed seems quite capable of being traced—for its antiquity is not so great. The colours which we now find in pebbles more or less opaque, though occasionally lively, exist under very different conditions from those by which they reside in the crystalline gems. It is necessary, however, to consider first, the origin and nature of *pebbles* as distinguished from casual fragments of stone.

This was long held to be one of Nature's riddles; but as soon as it was experimentally dealt with, it met with a solution in some of its most difficult points.

It was early decided that pebbles are *distinct formations*, complete in themselves, except in so far as they have been worn away by gradual attrition. The first difficulty was, how to account for the great hardness of many of our seaside specimens. Although for the most part inclosing an "organism," which must have been that of some zoophyte, they show no traces, in their present compact texture, of the soft and yielding consistency they must at that time have possessed. The matrix in which they lie, and from which they drop as the ripe nuts fall from a hazel-bush, is seldom or never of so hard a substance as are the pebbles themselves; in many cases the difference is as great as between the teeth and gums of a living mammal.

But the evidence of the inclosed organism is conclusive as to a past history; and in all mixed pebbles—and ninety-nine out of a hundred of ours are mixed—it is quite certain that an *impregnation*, or an actual *infiltration*, has taken place.

Either a fluid menstruum, usually siliceous, must have enveloped and saturated the animal form, or there was actual injection of chalcedony and limestone, in a soft state, into the tubes and cells of the skeleton first, and afterwards into the pores and crevices of the new stone. Frequently both processes have obtained. The only alternative, viz. that which suggests that composite pebbles, revealing in their structure distinct traces of animal or vegetable organization, *may have been* formed thus at first, like the coloured prints in a book, I consider to be untenable. One might as readily credit a spontaneous growth of almond-cakes or oyster-patties by some sudden spasmodic effort of our present sea and land.

Impregnation has, no doubt, been always going on.

A French *savant*, M. Reaumur, above a century ago, wrote as follows:—"By a coarse operation emery is reduced to powder and suspended in water for several days; but nature may go much further than this, for the particles which water detaches from hard stones, by simple attrition, are of an almost inconceivable degree of fineness. Water thus impregnated contributes to the formation of pebbles by petrifying the stone, as it were, a second time. Stones already formed, but having as yet a spongy texture, acquire a flinty hardness by impregnation with this crystalline fluid."[3]

From such a source, as he supposes, has arisen the close texture of Egyptian pebbles, coloured jaspers, and even agates. If he means to include the *homogeneous* agates, which are alike free from sparry crystals, metallic scum, and all traces of organized matter, I do not agree with him. But in the case of *mixed* chalcedonic pebbles, among which may be classed our pretty Isle of Wight specimens, no doubt such impregnation took place, and was followed by a further process of infiltration. For when the flint-nodules, *impregnated* as above, were still soft—soft

[3] "Mémoires de l'Académie des Sciences." Anno, 1721.

enough we know they were to take delicate impressions of the spines of the echi-nus—chalcedony, in a semi-fluid or viscous state, would pass through the pores of the flint, because the former is the finer substance of the two. And after such infil-tration, the entire lump would harden, resisting, for the most part, further change. And this would be the pebble as we now find it.

In the "Geological Museum," now open in Jermyn Street, there is a case (on the first floor) where the nature of infiltration is well shown in some jasper-agates. It will readily be seen here that each internal layer has been formed in succes-sion *from without*, the centre of the pebble filling up last. In another case (labelled "Silica"), on the opposite side of this room, are some "choanites" and "sponges" presented by the author.

Those latter will be found worthy of ten minutes' inspection, even as seen through the plate of glass which serves to protect them. They are selected, not as being the finest specimens in his cabinet, but as illustrating, each of them, a par-ticular animal, or a peculiar *position* of one in the fossil state. They are, however, very good specimens, much above the average; and if any one became missing, it could never be exactly replaced, though you should search the world over. Nature does not stamp the same "medal" *twice*. Any person who, after examining these, likes to start a theory of his own to account for their forms and colours, has my full permission to do so. But one word I would say in friendly warning, seeing that theories do abound. The Chinese, that ancient and wise people, have a theory that "asbestos" cloth may have been manufactured "of the hair of certain rats that lived in the flames of certain volcanoes." It were monstrous to doubt it.

In my beach-rambles I have often picked up and examined globes of sand-stone which were partially chalcedonized, and that by evident infiltration. The lapidaries call these "sand-agates," and reject them as unfit for the wheel; and so they are at present, but it would be instructive to meet with some of them twenty years hence, after they had undergone a more confirmed treatment at the hands of Dame Nature.

The well-known specimens of "petrified wood," common on our coasts, and occurring in some beautiful varieties of beech and acacia in the bays of South Dev-on, are a further example of infiltration; but the process here must have been some-what different.

The presence of metallic particles, what lapidaries term the "moss," in many of our agates, argues an impregnating fluid, thoroughly charged with mineral matter. In most cases this fluid was a ferruginous stream, such as may often be seen issuing from some hidden reservoir, and trickling down the face of a cliff of gault or greensand. This tinges everything that lies in its path with an indelible red stain; it then oozes on through the shingle, and reaches the verge of the sand. The nearest pool becomes saturated between tides, and the suspended crystals of salt, which are of a penetrating character (as the state of your beaching-boots will soon inform you), enter into chemical combination with the metallic rust, and help to conduct its particles into the heart of many a limestone pebble.

Dark inland pebbles, on the other hand, will discharge much of the oxide which they have imbibed, and may be observed to brighten and improve their complexions after a few months' sponging and tossing in the purer sea-water.

Some of our handsomer pebbles, when cut in two, reveal blotches of metal, which are glossy after polishing on the wheel. This metal is occasionally native iron. The darker varieties of "moss" contain a good deal of manganese, and some silver.

From what has been said about impregnation, we may readily conclude that the *colours* found in our agate-pebbles have been chemically inwrought, and that the symmetrical *patterns*, over which these colours are disposed, are not due to a succession of "layers," as in the ribbon-jaspers and Scotch pebbles, but must be referred to the organization of some extinct zoophyte, whose skeleton is preserved in the existing fossil.

The cause of translucency, or even *transparency*, in some pebbles is, no doubt, to be sought in their finer and more even texture, but especially in the latter quality. Sir Isaac Newton was of opinion that the opacity of certain substances is simply a result of their cross-grained composition. He held that in a transparent body the particles must be regularly and evenly disposed at equal intervals, so that a ray of light entering such a substance would pass steadily on, according to the known laws of gravity and motion, meeting with no obstruction beyond that of the homogeneous density of the medium which it had to traverse. Whereas, in opaque bodies, he supposes these constituent particles to be unevenly disposed, and that the ray which enters at the surface is, as it were, pushed about and thrust aside, and at length lost to sight.

The purity of spring-water, when undisturbed, illustrates this beautiful theory of the great philosopher; and, among solid substances, good glass, especially plate-glass which, after fusing in the furnace, has been carefully run and settled in the frames. It may also be simply shown in the act of wetting or steeping certain dry substances. Thus, stout writing-paper, owing to its cross texture, is opaque when dry; but if we immerse a sheet of it in oil, and then hold it up to the light, we find it has become transparent. For the smooth medium thus imbibed has entered its pores, and *equalized* the general texture.

We find also a striking argument for the probable truth of Newton's eagle-eyed conclusion in certain minerals which possess double refraction.

ICELAND SPAR, one of these, is perhaps the most perfectly transparent solid with which we are acquainted. Now, Iceland Spar *transmits* both the rays, as we see in the twofold image which is presented to us. But TOURMALINE is opaque, even the red specimens looking almost black; and we find that Tourmaline *absorbs* one of the two rays of polarized light. This singular stone polarizes a beam of light if it enter in any direction but that of the longer axis.

No solid with which we are acquainted is absolutely transparent. And there is no perfect *reflector* known. The best-polished surface *absorbs nearly one-half* of that quantity of light which strikes upon it.

As to our seaside pebbles, the most compact of them are even *visibly* porous.

Cut through the hardest jasper, and polish one face till it shines like a steel mirror, and then hold it so as to reflect the light, you will at once discern numerous minute specks or flaws on this surface, as if the point of a needle or graving-tool had been busily at work upon it. These specks have always been there. They are *air-holes*, and in the case of a fossil-pebble were, some of them, doubtless, connected with the position which the animal occupied just before he died. These air-holes must have traversed this hard stone in every direction, for, cut the pebble how you will, you meet with them. Others again, incomparably more numerous, but which can scarcely be discerned without the aid of a powerful microscope, are the true *pores* belonging to this apparently impenetrable substance.

I have found that the action of daylight tells even upon polished agates after a time. Some of the spots and markings in the finest stones grow fainter in the course of many years. Others appear to deepen. This is, no doubt, a change resulting from the oxygen, contained in a beam of light, acting readily upon carbon, silex, and other substances. And it furnishes an independent argument for concluding that the visible beauty of these formations, in perpetuating a coloured pattern, has been due rather to mixed *chemical* processes than to pure crystallography.

The fossil nodules, mostly siliceous, which we find upon a sea-beach, are *integral* specimens. It never but once happened to me to meet with what appeared to be *two* choanites in the same pebble, and I incline now to believe that this was in fact *one* choanite who had *divided* himself after the manner of the Luidia. But even supposing that here were, indeed, a pair of perfect animals, when it is considered that this was one single instance, occurring among more than a thousand picked up by me, in which the choanites were always solitary, such an exception may be almost said to prove the rule.

The late lamented Hugh Miller has stated the like fact concerning his Morayshire and Cromarty icthyolites. At pp. 194, 195, of his "Cruise of the Betsy," he remarks that "the limestone nodules take very generally the *form* of the fish which they inclose; they are stone coffins carefully moulded to express the outline of the corpses lying within." Then, after observing that the shape of the stone conforms to the attitude of the fish in every instance, he goes on to state that "the *size* of the fish regulates that of the nodule. The coffin is generally as good a fit in size as in form; and the bulk of the nodule bears almost always a definite proportion to the amount of animal round which it had formed."

He then gives, in his usual clear and graphic language, the following remarkable statement, the result of direct experiment.

"When a calcareous earth, mixed up with sand, clay, and other extraneous matters, was deposited on some of the commoner molluscs of our shores, it was universally found that the mass, incoherent everywhere else, had acquired a solidity wherever it had been penetrated by the animal matter of the molluscs. Each animal, in proportion to its size, was found to retain, as in the fossiliferous spindles of the Old Red Sandstone, its coherent nodule around it. Here the animal matter gave solidity to the lime in contact with it. But in the *natural* phenomenon of the icthyolite beds, there was yet a further point, for in these the animal matter must have possessed such *an affinity* for the mineral, as to form, in an argillaceous compound, a centre of attraction powerful enough to draw together the lime diffused throughout the mass."

Hugh Miller concludes the above passage by saying that "it still remains for the geologic chemist to discover on what principle masses of animal matter should form the attracting nuclei of limestone nodules."

Upon which last remark, I would suggest that the principle of the "law of definite proportions," established by Dalton, and fully borne out by all succeeding experiments in chemistry, appears to meet the case, geologically as well as chemically, in every respect. For such a law must of course obtain, whether the chemical ingredients be present in the forms of mineral, vegetable, or animal life; in fact, we have never known the law to vary. It holds for the compounded elements which we term AIR and WATER; it holds for the crystals; it holds for vascular and cellular tissues; why should it fail when the further instance of a mollusc or a zoophyte is in question?

I see no reason to doubt that the blood, muscle, membrane, or adipose matter of any animal have their due respective affinities for the different substances of the mineral creation.

No doubt, however, electricity, as an active cause, bears directly upon all such phenomena.

I remember many years since being greatly perplexed with the *discovery* (as I then thought it) that in certain parts of the Isle of Wight, and along the Dorsetshire coast, the rows of dark flints imbedded in the upper chalk of the cliff-ranges were all *broken*; broken *small*, so that if one were taken out of its chalk socket it crumbled in your hand. Often and often, I turned this fact over in my mind, and sought some apparent way to account for it. But I think I now see one way to account for it; and which may perhaps be judged worthy of acceptance, until a better solution is hit upon. In Dr. Hook's experiment with that once common household implement, the "flint and steel," he found that the *sparks* which "fly" upon collision taking place are minute spherules of metal. And further, that this metal was now not *steel* any longer, but iron; the fragments struck off *having lost their polarity* in the moment of contact with the flint.

This experiment shows that silex possesses some remarkable affinity for the magnetic fluid, since in this case it had robbed the steel of it, for those spherules would not answer to the magnet. May not, therefore, the rows of flints in the cliff have attracted the lightning in severe thunderstorms, and been shivered by the blow? This would not necessarily hurl the chalk down.

The Arabs have a proverb which says, "Under the lamp it is dark." Certainly, the chief mystery of every animated creature seems to reside *in its life*, and *Life* is like a burning lamp. Once the life extinct, we can anatomize and analyze, and arrive at many partial conclusions; but, meanwhile, the life has departed, and we do not know what that was. The nearest thing to its likeness is *expression*; and animals rank high in the scale of being, according as they command and impart expression. A dog or horse have a great deal; a bird some; a reptile or insect, absolutely none to our eyes.

With these latter, however, their *habits*, while alive, take the place of any more intelligent expression; especially some of their more excited movements. We know by the *sound*, when a snake is angry; and as to motion, the attack of a provoked hornet is a startling thing to witness. Ordinarily speaking, however, all their habits tend to concealment; no doubt, for the sake of safety. I believe no one ever sees the Sphynx-moth, called a "Death's-head" (*Acherontia Atropos*), on the wing. Yet we discern many others which, like it, fly in the hours of dusk. But this insect whirls through the air like a stone from a sling. It has never yet, in my experience as a naturalist, been my lot to meet with the burrowing insect called an Ant-lion. Only twice in my life have I encountered a slow-worm in the woods of Devon and the Isle of Wight. And to instance the case of a far commoner creature, how seldom does it happen to any one to take the large dragon-fly behind his green leaf!

And yet the problems of inanimate matter are perhaps the most difficult to solve. They have deeply exercised philosophers in every age of the world.

"All things considered," says Newton, "I think it probable that God, in the beginning, formed matter in solid, hard, impenetrable, movable particles; of such sizes and figures, and with such other properties as most conduced to the end for which He formed them. And that these primitive particles, being solids, are incomparably harder than any of the sensible porous bodies compounded of them; even so hard as never to wear, or break in pieces; no other power being able to divide what God made one in the first Creation. While these corpuscles remain entire, they may compose bodies of one and the same matter and texture in all ages; but should they wear away or break in pieces, the nature of things depending on them would be changed. Water and earth composed of old worn particles and fragments of particles, would not be of the same nature and texture now with water and earth composed of entire particles at the beginning. And therefore, that Nature may be lasting, the changes of corporeal things are to be placed only in the various separations and new associations of these permanent corpuscles."

So wrote the man, "Qui genus humanum ingenio superavit," who at the age of twenty-four had already invented the Fluxional Calculus, discovered the Decomposition of Light, and enunciated and proved, in a series of lemmas, the law of Universal Gravitation.

But soft! What is this which steals upon our senses through bobbing apple-blossoms at the open casement? It is the delicious, oystery smell of the Sea, as the afternoon tide runs in with a sound like that of fairy-bells upon the sand and rock and loose shingle.

That scent, and sight, and sound, are altogether irresistible. We throw aside Dr. Buckland and Hugh Miller, clap on a "wide-awake," rush down at once by the zig-zag path, and madly catch up a bunch of dripping tangle, and kick a hole in the moist sand, and presently find that our "Balmorals" are ancle-deep in the brine of a nimble wave.

And now I descry a faint purplish hue upon the red beach which lies at the foot of Culver. To-morrow, we will breakfast early, and hunt up a score of pebbles before the day grows hot.

CHAPTER VIII.

Pneumatics And Hydraulics.—Power Of The Wind.—Power Of The Waves, And Velocity Acquired By The Water.—Occasional Height Of Tides.—Tidal Wave.—Currents.—Tidal Phenomena On The Isle Of Wight.—Variations In The Coast-Line.

The elemental powers which are constantly at work in Nature all around us are upon a vast scale; a scale, in fact, which is immeasurable by us, except when observed under certain favourable circumstances. As our practical ideas of locomotion are immensely surpassed by the speed of the Earth's rotation on her axis, which is at the equator rather more than 1000 miles an hour, and yet far more surpassed by the rate at which she moves in her orbit round the Sun, which exceeds 20,000 miles an hour, so also our experience of a puff of wind on the hillside, or a dash of water in our face, when a slanting shower or the spray from a cataract salutes us, gives us no conception of the stupendous powers which are exerted when the sea, convulsed by a storm, rages along some unprotected coast.

Once in my life, I remember being knocked down by a blast of wind; and several times, as I suppose has happened to many persons, I have been *floored* by an unexpected billow while bathing. I always made this reflection at the moment: "Its power is not known: it has floored *me* easily enough, but perhaps it would have felled an ox."

But a wave impelled by the gale, which could take an ox off his legs, is nothing at all. What is the measure of force actually applied when a stout ship is shivered on a lee shore? No one knows. When the Clarendon, West-Indiaman, struck on that fatal "race" off Blackgang Chine, she went utterly to pieces *within seven minutes*! Yet, what a delicate thing is AIR! What a yielding thing is WATER! But then the air and water threw her *on a lee shore*.

Observations, carefully made at the right time and place, enlighten our ignorance upon all these matters. Air, delicate as it seems, when compressed, explodes in the roar of the thunder; and water is almost *incompressible*, and therefore its blow will knock down *anything*.

Let us speak first of the power of the WIND. The sands of the Desert, as has been said, are powdered quartz and quartz is a heavy substance; but when a strong wind ploughs the surface of the Desert at an angle, these Sands are lifted, and made to gyrate in spiral folds, and a huge column is formed—perhaps a hundred and fifty feet in height—and this monstrous trunk is presently carried along at whirlwind speed; and if it meet a party of travellers, they will be overwhelmed and buried.

It may be said that the sand is in very minute particles; and this is true. But if they are, therefore, the more easy to disturb and to catch up aloft, they are all the more difficult to bind together in a spiral form, and to hurry along the desert unbroken.

The power of the wind, especially when it comes in sudden gusts, and with a whirling motion, against timber-trees and the upper parts of buildings, is only too well known; on wide, exposed plains, its fury, in the early part of the year, sometimes brings destruction to everything within its range. But it is on the open sea, where the hurricane, once let loose from heaven, can sweep, unchecked, perhaps for hundreds of miles, that the latent forces of this unseen element are revealed in all their terror and majesty. For here the wind not only has free scope, but it also finds another element, fluid but incompressible, to obey its impulse and follow in the course which it takes. If the WIND be like a wild spirit, the WATER is a mighty, irresistible body, endowed with motion by the other, and capable of any work, from the drowning of a sea-gull to the wrecking of a stately ship.

The waves which roll in from the open sea, when the tide is *making,* are the most powerful. Fortunately, these do not reach the shore with quite the same force which they exert at the distance of some miles from it; but in exposed situations, their altitude and momentum are very great. The billows of the Atlantic which break on the western coast of Ireland run from thirty feet to fifty feet in height, and they arrive on that coast with an impetus which has (up to a certain point) been gaining strength, perhaps for half the time of a tide. It was this fearful onset, from a foe that never slumbered, which broke up and rendered useless the eastern extremity of the great Submarine Electric Telegraph Cable between Ireland and America.

It has now been decided that when stormy winds prevail long in one direction, another and peculiar force is given to the waves on which they operate. For the wind, by pressing long *on the side* of a wave, changes its form from that of an upright ridge (*i.e.* with *vertical* axis) to one which has a stoop or bend, sometimes of great inclination. Such a wave will come in upon the shore with far greater momentum, for its *velocity* has been enormously increased, while its *bulk* is nowise diminished. The speed which the water acquires under the influence of a prolonged sou'-wester on the coasts of Sussex and Hampshire, although there the gale has only traversed the breadth of the Channel, is very notable at times. At Lowestoft and at Peterhead, easterly points of land, the danger of the winds from the North Sea is proportionally far greater, as a larger body of water has been continuously acted upon. For it must always be remembered that a wave at sea is simply an *oscillation* of the water which rises and falls in that place. The identical wave does not pass on to the shore, though it appears to do so; it is the *motion* which is propagated, and, as it were, handed on, like an electric shock through the successive plates of a battery. Now this motion *increases* with the onward career of the gale. "Vires acquirit eundo." And when the mass of the fluid over which the wind is blowing has once been agitated, and the equilibrium of its upper layers been thoroughly disturbed, if the current of air not only continue, but wax stronger and stronger, there scarcely seems any limit to the momentum which the rolling wave may thus acquire. The arithmetic of a few facts upon this subject will, however, give us some idea of what that momentum must be on certain occasions.

It has been found, by experiment, that the velocity imparted to an incoming wave sometimes equals *seventy feet in a second of time*. Now this would give nearly one mile in a minute, if it were a projectile in free space; but the case here is, of course, different, since the wave is only a portion of the mass to which it belongs, and no individual wave travels very far. Still, the amount of rush and pressure which are exerted is altogether stupendous. Probably the *breakers*, which in an enduring storm rise against a lighthouse in the open sea, are among the strongest instances with which we are acquainted, and, next to these, I suppose, the billows off the Cape of Good Hope. But we may take those on the Irish coast, westward, as a well-known sample. I have often watched these—once in a gale of wind—and I have seen, in one as yet *unbroken* wave, a body of water which I should compute roughly at two hundred tons' weight, and as having, at the moment, a velocity of not less than sixty feet in the second, discharge itself upon the reef of rock. On such occasions, if you happen to stand on the shore, you will feel the ground *vibrate* apparently, though the sensation is probably electrical. But let any one consider the conditions of the wave cited above, and reckon what is *the force of the blow* at the moment of impact. I have no doubt it would knock down a good strong house, if delivered against the upright face of it.

One winter, when I was at Sandown, in the Isle of Wight, a spot frequently mentioned in these pages, I found I was *utterly unable* to traverse the bit of open road reaching from the Fort to the point where the lane turns off to Yaverland. Yet I had on at the time a stout Cording's waterproof, with sleeves, and a slouched hat tied under my chin, and held a powerful oaken stick in my hand to help me win my way. In fact, it was a trial of dynamics between me and the gale, whether this "body," acted on by a continuous muscular force, in nearly a right line, could be propelled through that resisting medium. I am sorry to say, it could *not*; but, in the attempt, it lost a neck-button from its coat, and was within an ace of being blown into the hedge, like a certain "man in Thessaly, wondrous wise," whom we have heard tell of in our childhood.

This storm held on for seven or eight hours, and as the sea rose many feet higher than usual under its influence, when the tide was well in, the strife of elements, and their combined assault upon the works of men's hands were truly grand. A breach was made through the solid causeway, compact long ago of clay, and gravel, and boulders of flint, and guarded seaward by dykes of timber. This breach was effected, I have no doubt, by the *stroke* of the wave, as the sword of Roland is said to have cut through the rock in the fight of Roncesvalles. A wall of stout masonry, not above five feet in height, and supported behind by earthworks, so as to resemble an escarpment, yielded before the weight of water, as a pane of plate-glass would give way at the charge of a locomotive engine. Some twenty yards' length of it was rent and thrown down. On examining this fragment afterwards, I found that the materials of the wall were solid blocks and angular bits of Wenlock limestone, and the cement used of the strongest. I believe, however, if it had been three times as strong a barrier, those waves would have levelled it.

I subsequently witnessed in that very neighbourhood a yet more furious tempest of wind, but had no opportunity of measuring its effects.

At Eastbourne, about two years age, I made acquaintance with a phenomenon in this line which was altogether new to my experience. We had a tremendous gale late in the autumn of 1857. I cannot exactly cite the quarter from which the wind blew, as I do not accurately know the points of the compass in their bearing on the town and bay; but the storm seemed to beat in from the south-east, as it faced the Marine Esplanade. I went forth, with perhaps a score of others, to gaze upon this magnificent "encountering shock" of earth, air, and water. We were all of us drenched to the skin by the dashing spray, and occasionally well-nigh swept off our legs by the gust; but we held on stoutly, *till* something saluted those nearest the beach, which rendered a retreat imperative. This was not the salt spray, nor the rattling hail neither, but a cloud of "skirmishers" in the shape of *pebbles and gravel* from the strand below. The sea had actually lifted the surface of the bed of beach, and whirled aloft some bushels of its solid contents. A coast-guard told me next day, that he saw a "flight" of pebbles, some of them as big as hens' eggs, at least thirty feet high in the air; and he added, that if the direction of the wind had changed a little, every pane of glass on the ground and first-floor of the Esplanade must have been shivered to atoms. As it was, of course much damage was done to windows and sashes; but not by the actual pebbles.

After this gale had subsided, I ascertained by personal inspection that some hundred tons' weight of solid shingle had been moved along the shore a distance of a quarter of a mile, in the space of a few hours. This was by the sidelong *drag* of the tide.

The village of Seaford, on this same line of coast, had long been in great danger of being swept away by the tide. Its outworks and very standing ground were perceptibly yielding, season after season.

About eight years ago, the inhabitants took the alarm, and drew up a remonstrance and petition, which were duly forwarded to the authorities. Proper officers were deputed to go down from London and report; and the result of their representations was that an able engineer from head-quarters was empowered "videre ne quid Respublica detrimenti capiat;" or, in the modern vernacular, to see that the Queen's lieges in Seaford suffered no wrong.

The remedy hit upon was simple and forcible; but, assuredly, it was an instance of what is called "Hobson's choice." The engineer said, "There is only one thing to be done; a solid breakwater must be formed west of the town, to break the rush of the sea and divert the course of the current. We will throw down a section of the chalk-cliff; that will make a mound, on which works of masonry, if necessary, can be erected afterwards."

This was done about two miles east of Newhaven, by firing an enormous charge of gunpowder in chambers drilled in the limestone. I went over from Brighton to witness it, and the sight was a striking one, as soon as the smoke and dust consequent on the explosion had cleared away. The operation, neatly conducted with an electric battery, proved successful; and its result, in the amount of chalk thrown down on the beach, was judged sufficient for the time.

But for this "pièce de resistance," Seaford might by this be in a fair way to furnish to future generations a gigantic specimen of a marine fossil.

The danger to Seaford arose not from storms or any casual visitations, but from the steady continuous action of the tide encroaching on the line of coast where the town stands. Whether the chalky breakwater above described will long suffice to counteract this elemental mischief may be doubted. If after a while it should prove inadequate, the operation will have to be repeated on a larger scale.

The height of the tides at sea is always known; and it varies little, depending upon astronomical causes, chiefly on the attraction exercised by the moon. But on shore, and inland in certain rivers, the case is widely different. Here the land itself, with its rocks and embankments, introduces artificial conditions which influence the local tides in an extraordinary manner. Thus, at Chepstow, by the Castle rock, the rise of a spring-tide is sometimes as much as *sixty feet perpendicular*. In Mount's Bay, Cornwall, it is fully *forty feet*; while in the north of Scotland on the coast, it does not average more than *ten*.

The tides have other variations beside this of local magnitude; and one of these, which takes place along our southern coast, is interesting and important, but was little known till the Admiralty surveyors went to work. It may be briefly described as follows.

As far as Scilly Isles and the Lizard's Point, the great tidal wave, which twice in every twenty-four hours flows in from the Atlantic, maintains pretty nearly the same level on the opposite shores; and this continues as far as Exmouth on the Devon coast, and St. Malo on that of Brittany. But when this great wave reaches the "Needles" which form the westerly point of the Isle of Wight, it divides into two very unequal portions. Of these, the southern and larger portion sweeps round the base of the island, passing Freshwater and Chale and Blackgang, and then rounding St. Catherine's promontory and running up to Dunnose with scarcely-diminished velocity; but the northern and smaller half enters the Solent, and owing, it is supposed, to the resistance of two shores (by *friction*) to its progress, advances but slowly, and does not arrive at Southampton Water before the other has made Dunnose and is filling Sandown Bay with a *back*-stream. This causes many differences in the amount of water at almost any given moment on the two sides of the Isle of Wight, north and south; and it gives rise to many curious varieties of tides, real or apparent. In one spot near Shanklin, my attention was frequently drawn to the fact of a partial tide suddenly *flowing*, for perhaps the space of an hour, during the time of ebb.

But this is not the only "water-company" here at work, nor is the above the only class of phenomena resulting from such agency. Beside the *tide-wave* there is the Ocean-*current*, which is quite a distinct thing. This current is due to a branch of the Gulf-stream; and it flows here from west to east, and from south to north. That is to say, after leaving Newfoundland, so much of it as actually reaches Britain would run up St. George's channel northwards, and up the English channel eastwards. And although the velocity of this current is but trifling, yet, owing to its unceasing action in one direction, its effects are remarkable. Moreover, it is not affected by the waves; whether the sea be sleeping in a calm or tossed with storms, the motion of the current is the same. The only change as yet observed is when its waters come to the surface and have their temperature lowered by a chill wind; for, this gulf-stream is itself a river of *warm water*, though its bottom and banks are of the same element *cold*.

Thus, between tide-waves and currents, changes are going on continually both in the volume and the physical constitution of the sea-water. And this, which might at first on a hasty glance look like casual and unimportant variation, when we come to consider it attentively, is found to be an arrangement fraught with wisdom and beneficence; for it insures continual variations of temperature, it attracts the purifying storms, restores the elastic spring of the atmosphere, and provides for weary man a healthy tonic breeze when he rambles along the beach or scales the face of the cliff. Without the motion imparted by the tides, the sea would probably become putrid. Now the tides depend upon the MOON; yet how seldom do we think of her invaluable services, when we gaze upon her pale face! An Italian once told me that he "loved the moon, adored the moon, never tired of looking at the moon." And the Padishah,[4] I have heard say, prefers "moon-faced" ladies for his soft companions; but I doubt whether either the pensive Italian or the glittering lord of the Bosphorus ever bestowed five minutes' thought on the mighty phenomenon of the TIDES.

There are other changes in what may be termed the tidal "high-water mark," due to the lapse of ages, which the sea has chronicled on the face of many a shore and cliff. It is well known that the coast-line, in this and other countries, has experienced many alterations as to its level. In some places, the ocean has apparently encroached upon the land; in many others, it has receded. One striking *index* of this latter, and which may be implicitly trusted as proving the fact, occurs where the waters have retired from their former level, and have left exposed to view the remains of an ancient *beach*. Thus, between Brighton and Ovedean, is found what goes by the name of the "Elephant-bed," because such fossil bones lie among the pebbles in this part of the cliff. The entire stratum undoubtedly was at one time a beach. Similar phenomena have been traced along the coast-lines of Arbroath and Cromarty shires.

Elsewhere, the sea has gained upon the land; undermining the friable sandstone cliffs, and, as it were, melting down the headlands, so as to change the outline of the coast. But in these latter instances, although "terra firma" has given way, it may be doubted whether the water itself has *risen*. Wherever it has indubitably done so, to the extent *e.g.* of submerging a village, I should apprehend the presence of volcanic agency.

---

[4] Sultan, "Father of the Faithful."

CHAPTER IX.

Geology Lies At The Bottom Of Earthly Things.

There seems to be no strong reason for doubting that every elementary substance of which we have any knowledge was *separated* at the first from that crude mass called "the earth," and which, along with "the heaven," God created in the beginning.

For, after that first assertion, we read of continued acts of "separation" by an Almighty *fiat*. Thus light is separated from darkness; waters from waters; water from earth. And by a like analogy we may suppose that the "firmament" or atmosphere, the lower part of which we breathe, was also eliminated in its component gases from the primitive shapeless bulk of matter.

If we take forty-five miles as the average height of this atmosphere above the surface of our globe, it is proportionately, like the thin rind upon an apple, or the bloom upon a peach; and may very well have come forth without any violence done to nature, or any waste of her resources, from the womb of the earth.

OXYGEN is a gas universally diffused, for without a large supply of it human beings and all the mammals, birds, and reptiles, would die. But oxygen is not a more *simple* element than CARBON or SILICON; and silicon is almost as universal, for it enters largely into the composition of all known rocks, excepting coal, limestone, and rock-salt.

Then, beside silicon and carbon, there is BORON, and many others, all of which will combine permanently with the principal gases, and which we may, therefore, fairly suppose did so in the beginning.

The oxygen which we breathe enters into composition with almost all known substances. It constitutes about one-fifth of the atmosphere, perhaps eight-ninths of the water, and helps chemically to compound all forms of earth and rock, and all the metals save five. Now this proclaims at once a *kindred* element, and does not look like a substance made independent of all the others at the first. Even *potash* and *soda*, which were long called "fixed alkalis," having resisted all attempts to decompose them, yielded at last to a galvanic process, and were found to be compound; oxygen, in both cases, having united itself intimately with a metallic base.

It is probable, therefore, that every substance which we can handle, or of which we can recognize the presence by a chemical test, was as truly a part of the chaotic Earth as was the clay or the granite. The gaseous atmosphere or "firmament" was eliminated by an act of creation from the torpid mass, the waters were drawn off and gathered together, the rocks consolidated, the metals precipitated, most of the crystals were probably perfected by galvanic operations. But there is nothing, from the crystallized blocks of the mountains to the salt held in solution in the sea, from the heavy rain-cloud which darkens the sky to the gossamer films of vapour which fringe the outermost edge of our atmosphere, which was not actually present as an existing particle when the Earth, "*Mother*-Earth," was made. And the EARTH, as we see, could very well furnish all these concomitant and subsidiary elements, without being impoverished at all. Indeed, as far as experiments and investigations proper have gone, we have as yet only touched the *surface* of our globe. The deepest mine which men have ever sunk into the bowels of the earth bears in its depth the same proportion to the earth's radius, as a *line*, or one-eighth of an inch, bears to the height of the monument at London Bridge. Some of the exposed strata of rock, allowing for the *tilting-up* of their edges, certainly reveal a deeper sample than this; but still, it is like the prick of a pin on the hide of a rhinoceros; it does not even penetrate the creature's folded hide; but of what lies *within* it tells us nothing.

Mr. Baily's successful attempt at "weighing the Earth," which, with consummate patience and skill, he brought to a termination in his house in Tavistock Square, has established the physical fact that this planet is not "a hollow sphere," as some persons had supposed, but *solid*, probably to its very centre, and of considerable specific gravity. Here is a mighty mass of Matter; it may be, as Dr. Whewell has argued, the *greatest mass* (in solid contents) in our Solar system. What do we know of its interior construction, economy, and arrangement? Nothing, except that all obey the *laws* established by the Creator, whether those of gravitation, cohesive attraction, or chemical combination. The "statistical report" is wanting, especially in such departments as those of the Trap and Volcanic lavas.

The amount of SALT in the ocean is a circumstance which geologists have not, I think, sufficiently considered. This salt, if precipitated, would, it is believed, yield a solid range of mountains equal to that of the entire Himalaya. Now, mighty rivers run into the sea, as those of La Plata, Amazons, Mississippi, St. Lawrence, Orinoko, and many others; but *none of these rivers are salt*; they are enormous bodies of fresh water. How, then, does the ocean maintain everywhere its saline character unimpaired? not to ask, whence did so much salt come? There is, indeed, a hill of salt in Spain, and there are mines of rock-salt in Poland and Hungary; but the *effort* which was made, so to speak, when the proportion now existing in the ocean was squeezed out of the huge terrestrial sponge, must far have exceeded that of any subsequent addition or contribution.

The ocean itself, as *girding* this globe, is a prodigious mass of water, indeed. But, if we remember the size of the globe itself, as given by its diameter, Ocean appears like a rain-pond. For, if we call its average depth five miles — and it cannot be more than this, may be much less — such a watery envelope would be represented in its proportion to the solid sphere which it surrounds, by a coating of varnish of the thickness of this paper lying on a globe which is twenty feet in diameter.

The COAL-MEASURES are also a remarkable item in the list of substances met with in beds and ranges, and claiming to be considered as rocks — for the coal is a *petrifaction*, perhaps a partial *crystallization* — but due to the deposition of vegetable matter, subjected afterwards to enormous hydrostatic pressure. We have no true estimate of their amount as yet, for fresh mines are being continually discovered; but here seems to consist the chief proportion of CARBON, as developed in this part of the visible creation.

CHALK constitutes one-eighth part of the crust of this globe. The mass of this must have been thrown up *at once*, when the atmosphere was eliminated, and the waters separated, and the sphere became condensed and solid. It is a "carbonate of lime" in its chemical description, and there is abundance of the limestone-rock all over the world; but it is very difficult to extract from this, in any quantity, the soft, friable, slightly unctuous mineral which goes by the name of chalk (Latin, "calx").

The METALS exist in the rocks, generally some distance down in the earth's interior. Their amount, in some instances, is very great indeed. The only one which is rarely met with on our globe, and then in small grains, is the NATIVE IRON, *i.e.* iron in its *pure* state. This want, however, could not have been recognized if we had not happened to find the substance in question in those blocks of meteoric iron, and in the small meteoric stones which have, from time to time, fallen to the surface of the earth during atmospheric storms, and, probably, owing to astronomical disturbances. For, of the iron, in its mixed conditions of different ores, we have an abundant supply. It is a singular fact, that the *native gold*, both in pure grains or flakes, and in solid "nuggets," appears now to be the more prevalent form of that ponderous and imperishable substance. The ranges of quartz-rock in many chains of mountains are certainly thickly sown with seeds and nodules of fine gold.

I may mention here, as an interesting circumstance to mineralogists and collectors of fossil pebbles, that the "native iron," which is so scarce on the surface of the ground, is met with, from time to time, along with *manganese*, in the heart of siliceous pebbles. But I have never seen it where there is not the presence of some animal organism.

Now all these *solids*, gold, iron, chalk, coal, salt—and I select incongruous items on purpose—are as truly portions of the earth's mass as are the granite and slate, the sandstones and clays. And *so are also* the flowing waters and the suspended clouds; and *so are also* the component elements of these—oxygen, hydrogen, and the rest. It is, all of it, MATTER; it all obeys the law of gravitation, and it needs just as much to be taken into account, in a geological scheme or system, as the upheaving of granite rocks, the flux of lavas, or the icthyolite beds amidst the semi-crystalline ranges of the Old Red Sandstone formation.

Consequently I have found, as has partly I trust been shown in the preceding pages, that even to handle and pronounce upon a pebble, such an one as we can pick up on the sea-shore any summer's day, *all these elements* must be taken into the account.

In a "moss-agate," for instance, I discover—

First. A sandy cuticle, which we will rub off, as it is like the dust on your drawing-room table, a real thing, but out of place.

Secondly. Siliceous matter.

Thirdly. A purer form of this in chalcedony.

Fourthly. The "moss," which is a metal proper, tinged by *oxidation*. [N.B. Here is oxygen inside a pebble!]

Fifthly. The whole or a part of some extinct animal's petrified structure.

Sixthly. One thing more, *sometimes* a drop of water.

Seventhly and lastly. Air, permeating the stone in fine pores and channels, one of the largest of which was in some cases a breathing-hole used by the zoophyte before he drew his last gasp.

Who shall therefore say that a PEBBLE from the sea-shore has nothing remarkable in it or about it, merely because the passing schoolboy can pick it up if he pleases, and, without looking at it for a moment, can fling it at the head of a gull, or dash it to atoms against a larger stone? It is a microcosm in itself; and if it lead us on to further inquiry and patient thought, it will amply repay our trouble, though we *have* loitered away a summer morn or an autumn evening among the Pebbles of the Beach.

CHAPTER X.

Guesses At The Probable Features Of A Past Scene.—What Was The Earth's *Commencing* Rotation.—Law Of The Tides At Such A Time, And Corresponding Action Of The Sea Upon The Land.—Original Character Of These Fossil Forms.

Whatever may be thought of the apparent scope or tendency of some of the geological theories which are rife in the present day, no person who has really considered the subject in its principal bearings has any doubt that the surface of our globe, both as to the land and the water, was once very different from what it is now. All sound argument must allow this to be possible: all careful investigation pronounces it to be the fact.

There are potent "menstrua" and mighty furnaces in the laboratory of Nature; and while the actual *amount* of MATTER in existence has, we suppose, never varied since the original creative act which gave birth to "the heaven and the earth," the conditions, sensible or chemical, under which that matter presents itself have altered from century to century and from moment to moment.

It is impossible to deem otherwise, if we believe that the "laws" which are established now in this creation were established then.

Let us take, for instance, the law of Universal Gravitation, that law by which the heavenly bodies move in their orbits, and a seconds' pendulum measures *time* at the surface of our Earth.

We feel assured that this law has never varied; but it is almost equally certain that it must have *begun* to operate *under different conditions* from those which obtain at this day.

Newton, who always maintained that the commencing axillary movement of the Earth was due to an impulse from the hand of the Creator when He launched it into space,[5] and who declared that without this he could in no way account for its having rotatory motion at all, was nevertheless willing to allow that the time occupied in the first rotation may not have been, as now, twenty-four hours, but possibly an entire year. And as the Earth, once projected, would fulfil, according to the Law of Gravity, the conditions of "a body falling through space," we find that taking the nth term (2n-1) of the arithmetical series ... 1. 3. 5. 7. ... (2n-1) ... where the successive terms represent the *velocities* acquired in successive annual periods, about 183 such terms must elapse before 365 rotations would be accomplished within the space of one year, or, in other words, the rotation of the earth be *diurnal,*

---

[5] See his Letter to Dr. Burnet.

as at present. This point of speed once reached, the centrifugal and centripetal forces balanced one another, and the axillary movement became steady.

Of course, in the above hypothesis, the assuming of one year for the first period is arbitrary; it may have been more, it may have been less; but within certain limits there is no absolute reason for supposing that it was a span of twenty-four hours, as now. *Revelation* nowhere asserts such a dogma.

Supposing, then, amid confessed ignorance as to how the fact stood, that the Earth's early times of rotation were far longer than now, and taking, say, *one year* as the first period, how would this influence other features in a scene which has now long gone by? How would it influence the Tides of the Ocean?

Assuming a correspondingly slow motion of the Moon in her orbit, we should, it seems, have in the lapse of the first year *one* rotation of the Earth and *two Tides* of the Ocean; in the second year, *three* rotations of the Earth and *six Tides* answering to them; and so on.

Now these flowing Tides would, in fact, have been vast *inundations*, the sea rising steadily for many months together; and in like manner the prolonged ebb which followed upon each flood would have given rise to a *subsidence* and to the deposition of such particles of mud, lime, silex, &c., as the waters then held in solution after their visiting the higher land.

And thus we see at once, without going further, that the agency exerted by the ocean in ancient times may probably have been different from and greater than any with which we are now actually acquainted, save in the account given to us, in the inspired Record, of the Deluge in the days of Noah.

When such mighty agencies were in operation, it is not unreasonable to suppose that great changes took place in the way of partial extinction of animal life, and the substitution of new forms to fill up the apparent gap left by the perishing creatures. The contents of the upper chalk, of the greensand, gault, and sandstones, as has already been observed, point to such revolutions and cycles in the history of animated nature.

MAN, who is himself an evident exception to all this, may perhaps, *as* an exception, be said to prove that such had been the rule. For, it must be remembered, *all* the creatures were pre-Adamitic. Not only those vast saurians and mammals, whose fossil remains we have exhumed, and cannot contemplate without wonder, were prior to our race in their actual possession of the domain of the earth's surface, but every bird, reptile, fish, and zoophyte were certainly made before the man. Now, as the man came last in order, but first in dignity, created in his Maker's image, and endowed with dominion over all the works of His hand, there was no longer place or argument for extinction, substitution, or change, among the creatures. *Death* was excluded, and could not enter into the world, unless by a moral delinquency of the chiefest creature.

But already it is probable, although he perhaps knew it not, Man stood, even in the day of his innocence and happiness, in the midst indeed of a blooming creation, but upon the crust of a fossil world.

Moreover, we know from Scripture that minerals find metals abounded; so that the presence of these near the Earth's surface ought not to be referred (as by some they have been) to the epoch of Noah's Deluge. "Gold and bdellium and the onyx-stone" were already in the "land of Havilah;" and Tubal-Cain was "an artificer in brass and iron," which must therefore have been exposed in veins of the upper rocks.

And if the Earth already showed her nuggets of gold and lumps of onyx, it is not likely that she was deficient in beautiful fossils. Certainly, such substances as manganese and native iron did not find their way into the heart of siliceous pebbles in *modern* times.

But it by no means follows from the above, that all these fossil forms belonged to creatures whose species are wholly extinct. Indeed, in some instances, this cannot be allowed for a moment. Probably, the "choanite" is extinct. I doubt whether the "ventriculite" be so. Dr. Mantell was pleased to assume that this last-mentioned zoophyte *grew*—like one of the "mushroom" class of plants—rooted to one spot. There is no evidence for this whatever, and my impression leans another way. In the class of marine "Acrita" termed *Acalephæ*, there is a family which bears the name of *Rhizostoma*. These animals are composed mainly of a large, mushroom-shaped, gelatinous disc, and of a supporting central pedicle. In structure and function this latter organ much resembles the root of a plant; and, no doubt, it absorbs nourishment by seizing upon the minute animalculæ which abound in the water where the "rhizostome" is floating about. The entire animal is very like a toadstool, or "agaricus," in form; but for all that, he does not grow, as a plant, in one spot, but is always sailing about by means of his large disc above-mentioned. His general appearance closely resembles that of a "ventriculite" from the flint; and I think it a probable guess to assign them to the same species.

The "branching alcyonite," on the other hand, did, I imagine, grow, like a Coral or "Encrinite," in one place; and I suppose it was much like an ice-plant in form, but that it had the power of drawing back at will all its branches and suckers—which were, in fact, the creature's arms and tentacles—into the root or bulb which formed its base. The "sponges," a large family, we know are not extinct; and the conformation of the living individuals fully bears out all the marvels which have been predicated of the fossil animal. I have handled these creatures, fresh from the sea, at Brighton; and I had an unpleasant consciousness, while holding one in my hand, that the round, greenish, jelly-like bush was only his *house*; but that the gummy fluid, which held possession of it after squirting out the sea-water, was the individual himself, though independent of the accessories of bone and muscle. This kind of "sponge" floats about: there is another, well-known among submerged rocks, which is fixed, and grows. I believe the fossil specimens to have belonged to the migratory class.

I once picked up, near Hove, a pebble which contained a fragment of the lungs of a tortoise; the elevated ridges, and depressed pulmonary cells, appeared to have left their peculiar structure in the flint. I have mislaid this specimen, and so cannot give a sketch of it here.

The "holothurida," or "sea-cucumber," is found occasionally in the Isle of Wight, beautifully fossilized. A specimen was shown to me, from the beach near Chale; it was a *variety*, but unquestionably of this class of animals.

A fossil, figured in one of the "chromo-plates" to this volume, has obtained the name of a "troglodyte;" I do not know why. I suspect it was one of the "asteridæ" when alive, but that in the death-struggle, the long arms collapsed and twisted together.

I have frequently found what I believe to have been some such organ as the stomach of a star-fish in the centre of hard pebbles. The form was always pentagonal, like the corolla of certain flowers pressed flat.

The "myriapod," depicted in Plate V., I should at once conclude to have been a marine insect, answering to some of our "scolopendridæ," if the *head* were not lacking. But since the back or spine in this fossil is all in one piece, and there are no lateral plates or divisions for the several pairs of feelers, it may be, in accordance with such a frame, that the head fitted on without the apparent juncture of a neck; and if so, this may be an entire insect. It is right however to add, that I feel some doubts about this.

The Plate VI. contains a figure of a fossil to which I have attached the name of "Spindle-choanite." In fact, the specimen, until cut in twain, was fusiform; and, I have no doubt, a complete animal.

The choanite from Eastbourne, *uncut*, Pl. IV., portrays a creature who, I think, expired in a vehement struggle. This would keep him on the surface of the siliceous flood; and it would harden, for the most part, beneath him, leaving his limbs sprawled out on the top of the pebble, as now seen. The beautiful "pyriform" specimen, Pl. VI., shows a similar struggle, amidst liquid agate and manganese. In the large Actinia, or "star-choanite," portrayed in the frontispiece, the happy animal died quite quietly; and that is why his fossil mummy is of such a noble size and development.

The figure called "Nondescript," Pl. IV., I do not think represents any complete animal, but a part of the organization of one. I have traced its likeness, in the "Animal Kingdom," by T. Rymer Jones; and think it not improvable that a woodcut at p. 364 of that most interesting volume depicts, in the *arms* of one of the "Brachiopoda," in that "calcareous loop" with its curved "crura," the exact structure which, in the living animal, served to open or shut the bivalve shell, but in the fossil specimen looks like a miniature-painting on ivory, to perpetuate some curious organism in the liveliest colours of the pallet. "Terebratula Chilensis" is Mr. Rymer Jones's animal; and in the very vicinity where I found my fossil in question, I have since obtained, in a flint pebble from the cliff, a beautifully petrified "Pecten" (same species of shell) in agate.

I might multiply remarks and instances such as these; but it is not needed. Every one will form their own opinion, in collecting fossil pebbles, as to what the *originals* were; but I think they will generally agree with me, that while our living individuals differ in many points, still we have mostly types of the same species actually existing in our present seas.

What all connoisseurs, and even amateurs, should do, is to preserve every remarkable specimen they may obtain, along with an accurate note of where they got it. It is a great pity that, among so many persons who have at once discernment and a liking for this branch of Mineralogy, there are so few who will be at the trouble of what lawyers call "taking minutes" of the evidence which comes before them in the course of the pursuit. I have looked over several fair enough collections made by amateurs; but, with one exception, I never knew an instance in which the collector could tell me, with any degree of certainty, from what particular beach he had gathered this or that specimen. Of course, when a scientific investigation is proposed, such omission renders the most valuable part of the collection comparatively useless. For the locality is all-important. A pebble may thus be always traced to its geological "home" or birthplace. It can only have had one of three sources: either it was first dropped *here*, where you find it; or, it is a "travelled" pebble, having come round from some distant beach; or, it is an offering from the deep sea. The latter case is of exceedingly rare occurrence, and need hardly be taken into account; nevertheless, for truth's sake, and to include all possible varieties, I cite it here.

A genuine deep-sea pebble is a waif which I have only twice, as yet, met with in my sea-side experience. One of these was, I think, fished up in a dredging-net; the other was thrown on the beach during a storm. In both instances, the fossil was a "choanite," and bore the marks of having been washed out of a limestone rock. The age of such fossil could not have been less than 4000 years, the date of Noah's deluge, for obvious reasons; but it may have been considerably greater.

If there were any known method of softening these pebbles to a consistency like that of melted glue, we might learn something concerning their past history; but I am not aware of the existence of any such process at present. *Silex*, once hardened into the condition in which we meet with it, is a most intractable substance to deal with; and the very fact of its having at one time been viscous, renders it highly improbable that, in the course of nature, its texture will ever assume a plastic character again. Moreover, while the majority of our flints are concretions, there are some which are semi-crystalline. These latter, it is evident, cannot now advance to the stage of perfect crystals; but neither can they retrograde: the next change which awaits so hard a substance must be, to crumble away.

Sometimes, on our beaches, hollow globes of flint are picked up, which, when you break them open, are found to be full of a white, powdery substance, like the chemist's magnesia. This is much the same as the "rock-milk" of mineralogists, a very fine deposition of lime, occasionally met with in beds of the chalk strata, and considered to be a result of some filtering process, when the water, after being long pent up, had escaped by small crevices. But in the flint-globes no such straining can have taken place; and in sundry specimens which I have examined, I generally came to the conclusion that some conchiferous animal had been inclosed in the nodule, and his shell had afterwards broken and been pulverized. Sometimes I found within the "rock-milk" a dark-brown, carbonaceous spot; this would be the creature's body, as Dr. Mantell supposes of his "molluskite." The cavity is never quite filled up. I imagine many of these "powder-horns" to be altogether modern in their date.

A much more curious phenomenon than the above occurs from time to time in solid agate-pebbles, which, when picked up, are found to contain in a central chamber, visible to the eye, a drop of water. These are scarce, and fetch a high price at the museums. As much as twenty pounds has been more than once given for an undoubted specimen. How the above-mentioned singular effect was arrived at, has, I believe, never been explained; at least, not so as to satisfy a close reasoner. The same *lusus naturæ* is met with, I think more frequently, in rock-crystal and in the dark fluor-spars.

The last apparent animal organism which I shall notice, as having more than once occurred to me as a *possible* explanation of the patterns disclosed in certain sections of agate-pebbles, is that of some creature's "ovary." The eggs of the whelk, and those of the cuttle-fish, are deposited in clusters, like some of the "grapes" on sea-weed. I have found, in some of the Isle of Wight pebbles, a conformation closely resembling this, but on a much smaller scale. That the original substance was part of an animal, I have no doubt whatever; and the arrangement of the lobes or spherules, composing the mass, more resembles that of the spawn of the above-named fishes than anything else which I am acquainted with. Still, resemblance is not identity. The thing may be only a curious coincidence, resulting from some unexplained freak of nature; as when she carved the profile of Napoleon's face in the outline of a part of Mont Blanc, and also (quite as distinctly) in that of a hill which overlooks the town of Belfast.

REFERENCES TO THE CHROMO-PLATES.

In the frontispiece is represented the polished section of a pebble which the author picked up on the beach at Bonchurch, in the Isle of Wight. This is an unusually large and perfect specimen, the body of the Choanite lying nearly central. The pebble contains one or two blotches of *native iron*. The "cuticle" is uninjured.

PLATE I.

Fig. 1. This is a slice from a lump of "conglomerate" found on the beach at Sidmouth. The *white* parts are sections of the nodules of quartz, the *red* and *yellow* are jasper.

Fig. 2. A section of an "Alcyonite" from the bay of Sandown. It formed part of a large pebble.

PLATE II.

Fig. 1. This is, I have no doubt, a fossil "Actinia." I have often looked on the exact living resemblance of it at the fish-house in the "Zoological Gardens," where it is fond of clinging like a limpet against the vertical pane of glass in an aquarium. — (*Found in Sandown Bay.*)

Fig. 2. I suppose this half of a pebble to represent the internal structure of some creature which dwelt in a bivalve shell. — (*Sandown Bay.*)

PLATE III.

Fig. 1. This sponge, a faultless specimen of the kind, is from the Brighton beach. It was the first pebble I ever picked up there.

Fig. 2. The body and arms of this Choanite are in white Agate; the remainder of the stone is a dark moss, formed chiefly of *Manganese*, and surrounded by a yellow flinty rim. — (*Brighton Beach.*)

PLATE IV.

Fig. 1. A Nondescript: but, probably, the creature was of the vermicular kind. — (*Brighton Beach.*)

Fig. 2. A handsome Choanite, uncut, but polished over, so as to show the points of some of the feelers. — (*Found at Eastbourne.*)

PLATE V.

Fig. 1. An "Eyed" Jasper, from the beach near Shoreham. It contains yellow "oxide of iron," and some dark green flint.

Fig. 2. A "Myriapod." This fossil is a very handsome one, and I have another, closely resembling it, from the same locality, in Sandown Bay. The dark, reddish spot, is of the nature of *Molluskite*.

PLATE VI.

Fig. 1. A Spindle, or *Fusiform* Choanite, when the pebble was entire. I do not possess a more perfect specimen: the Chalcedony is remarkably fine, and the "oxide" of a rich tint. — (*Sandown Bay.*)

Fig. 2. Pyriform Choanite, uncut. Here, again, the creature lies over the surface; and, as I conceive, from the position which it occupies, was swimming for its life. — (*Found at Rottingdean.*)

PLATE VII.

Fig. 1. An "Asterid." This is different from all the other creatures in these fossils. Its position in the heart of a solid limestone pebble is singular. — (*Beach at Hove.*)

Fig. 2. — "Terebratula." The entire pebble was formed inside of a "Pecten"-shell, and *inside the pebble* lies this formation, which was a living organism connected with the hinge. — (*Beach near Luccombe.*)

EXPLANATION OF CERTAIN TERMS USED IN THIS VOLUME.

"Hard:" that which will cut or scratch other substances.

Example. — A diamond will cut glass.

"Tough:" tenacious; whose particles are difficult to separate from one another.

Example. — Jasper, though not nearly so *hard*, is more "tenacious" than diamond.

"Transparent:" through which we can see objects.

"Translucent:" through which we see light.

"Opaque:" through which we can see nothing.

"Brilliant:" reflecting or refracting the light in rays and flashes.

"Vitreous:" glassy in its texture.

"Conchoidal:" convex, like the outside of a shell.

"Momentum:" the result of the combined weight and velocity of a body in motion.

GEOLOGICAL STRATA REFERRED TO IN THIS VOLUME, ARRANGED IN THEIR DESCENDING ORDER.

Names of Strata.	*Mineral products.*
Lava	Pumice-stone, which is its froth or scum.
Granite	Corundum, sapphires, felspar.
Old Red Sandstone	Fossil ichthyolites.
Carboniferous Series	Coal, lignites, jet.
New Red Sandstone	Jaspers, fossilized wood.
Lias of the Oolite	Fossil "Saurians."
Wealden	Fossil reptiles and mammals.
Gault	
Greensand	Crystals, flint-nodules, choanites, echini, cray-fish, lignites.
Chalk	
Tertiary Sands	
Clays	Amber, carnelians, fossil shells.

GLOSSARY OF GASES, MINERALS, GEMS, CRYSTALS, AND FOSSILS, NAMED IN THIS VOLUME.

Simple Substances.

Oxygen: a gas.	
Silicon	
Boron	Elements allied to the gases.
Carbon	
Alumina	
Magnesia	Soluble in acids.
Sulphur	
Manganese	
Iron	Metals.
Gold	
"Steel" is Iron which has been "carbonized."	

Derived or Compounded.

Diamond	
Sapphire	
Ruby	
Emerald	
Topaz	
Amethyst	
Aquamarine	These are termed "gems" or "precious stones."
Garnet	
Pearl	
Turquoise	
Jargoon	
Onyx	
Sardonyx	
Quartz	
Rock-crystal	
Tourmaline	These are crystallizations.
Spar	
Asbestos	
Jasper	
Chalcedony	
Agate	
Carnelian	
Flint	These are concretions.
Bloodstone	
Moss-agate	
Weed-agate	
Mocha-stone	
Jet	
Amber	
Clay	These are exudations, depositions, or fossils.
Chalk	
Coal	
Rock-salt: a chloride of sodium.	
Kaolin: this is from "felspar."	

Echinus	
Ammonite	
Alcyonite	
Troglodyte	
Choanite	These are fossil animals, for the most part belonging to extinct species of "zoophytes."
Ventriculite	
Icthyolite	
Cray-fish	
Sharks' teeth	
Sand of the Desert: this is powdered quartz.	
Sand of the Sea-shore: sandstone dust, mixed and penetrated with salt.	
Sea-weed: a compound of vegetable carbon, salt, and siliceous particles.	

MINERALOGICAL AND CHEMICAL KEY TO THE GLOSSARY.

Oxygen.	This is a pure gas, and the most universally diffused substance in nature.
Silicon	
Carbon	These elements will combine permanently with the principal gases.
Boron	
Alumina	
Magnesia	These elements, when dissolved in certain acids, yield a colourless solution.
Sulphur	
Manganese	
Iron	These elements, dissolved in acids, yield a coloured solution.
Gold	
Delessite is a chloride of iron.	

Silica comprises only 2 species:—1. Quartz; 2. Opal.

From *Quartz*, come

Rock-crystal	
Cairn-gorums	which are "vitreous,"
Sand	
and	

Chalcedony	
Agate	
Onyx	
Carnelian	which are "concretions."
Sard	
Bloodstone	
Jasper	
Flints	

Opal is a "hydrate" of SILICA, containing 7 per cent. of water.

Precious, or "noble."
Iridescent.
Common (colourless).

"Silicates."

Of Alumina	Feldspar.
	Kaolin.
Of do. with Glucina	Emerald.
	Beryl
	Garnet
	Pyrope, or "precious garnet."
Of Magnesia	Chrysolite.
	Meerschaum.
	Asbest.
Of do. with Fluorine	Topaz.
	Tourmaline.
Of do. with Zircon	Jargoon, or "Hyacinth."

The colour of "emerald" is due to the oxide of chrome; that of "beryl" to the oxide of iron.

CARBON comprises only 2 species:—1. Diamond; 2. Graphite.

"Diamond" is a perfect crystallization, and is the hardest substance known.

"Graphite" (sometimes called "Black-lead") is a carbonate of iron. The iron, however, enters in very small quantities, and is now supposed to be *accidental.* "Graphite" is a concretion, and is never met with in the form of crystals.

"Carburets."

Of Hydrogen	Bitumen.
	Coal.

Bituminized wood is Lignite, which, when very compact, is Jet.

"Amber" is a vegetable resin. It is obtained from rivers in Sicily, and from mines in Russia.

	Limestone.
"Carbonates" of Lime	Calc-spar.
	Chalk.

The "marbles" of Paros and Carrara are *crystalline* Limestone: those of Siena are *compact* Limestone.

ALUMINA comprises 2 species:—1. Corundum; 2. Sapphire.

1. "Corundum" is a very hard crystal, the common type of all the Sapphires.

2. Oriental "Sapphires." These embrace 6 kinds.

> The blue Sapphire.
> " red or "Ruby."
> " green or Emerald.
> " purple or Amethyst.
> " golden or Topaz.
> " sea-green or Aquamarine.

These are pure Alumina.

"Emery" is a coarse variety of Corundum.

"Aluminates."

	Spinel ruby.
Of Magnesia	Balas do.
A Phosphate of Alumina and Magnesia	Turquoise.

The "Spinel" ruby is scarlet; the "Balas," of a faint pink. The "Oriental" ruby alone has the "pigeon's blood" hue.

LOCALITIES OF SUNDRY "FOSSILS."

(Name of Fossil.)		(Proper locality.)
Echini	Galerites Albogalerus	Found in the Upper Chalk, and, occasionally, in the Greensand.
	Ananchytes Ovatus	
	Cor Anguinum	
Ammonite		The Lias.
Alcyonite		The Chalk.
Troglodyte		
Choanite		
Ventriculite		
Ichthyolite		The Old Red Sandstone.
Cray-fish		The Gault.

Shark's teeth	The Tertiary Strata.

PLATE I.

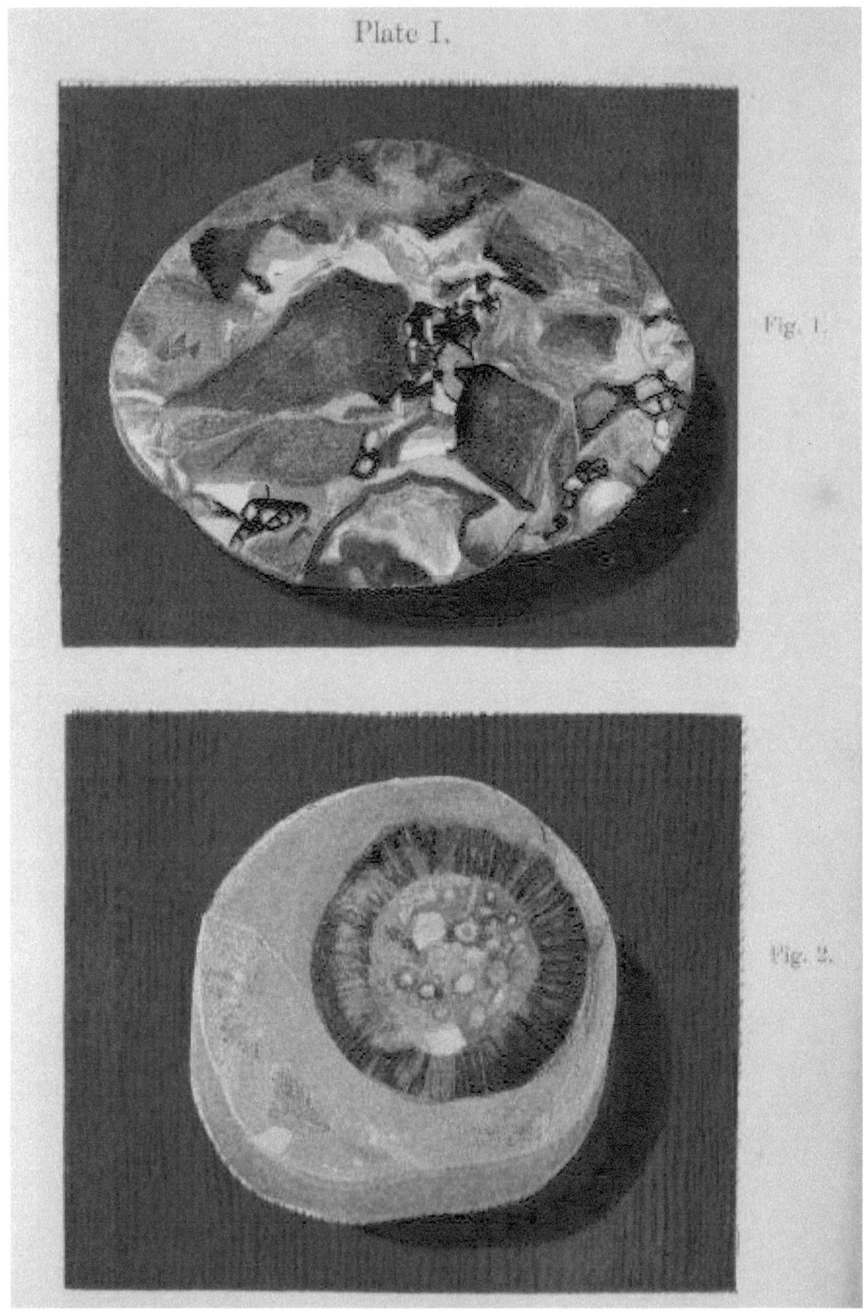

FIG. 1. FIG. 2.

PLATE II.

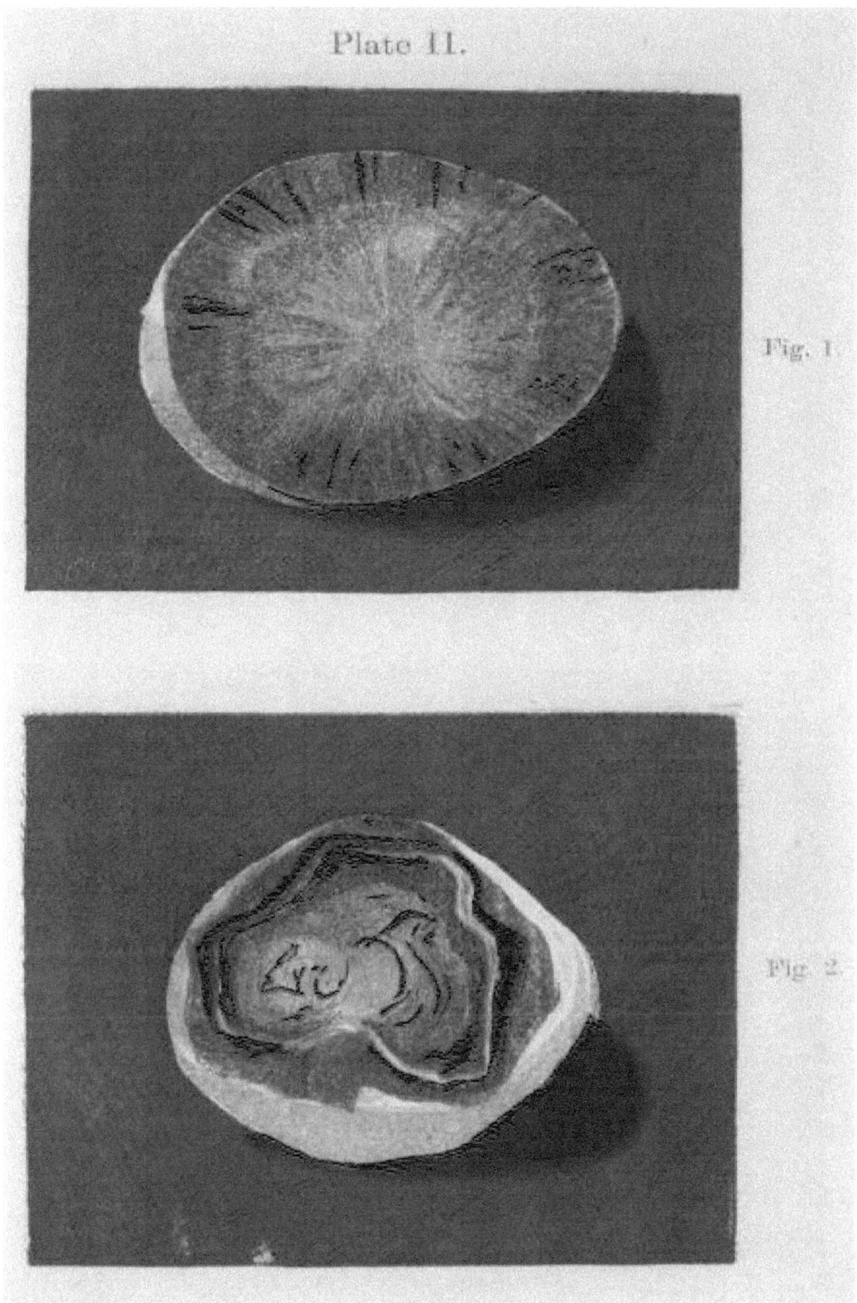

FIG. 1. FIG. 2.

PLATE III.

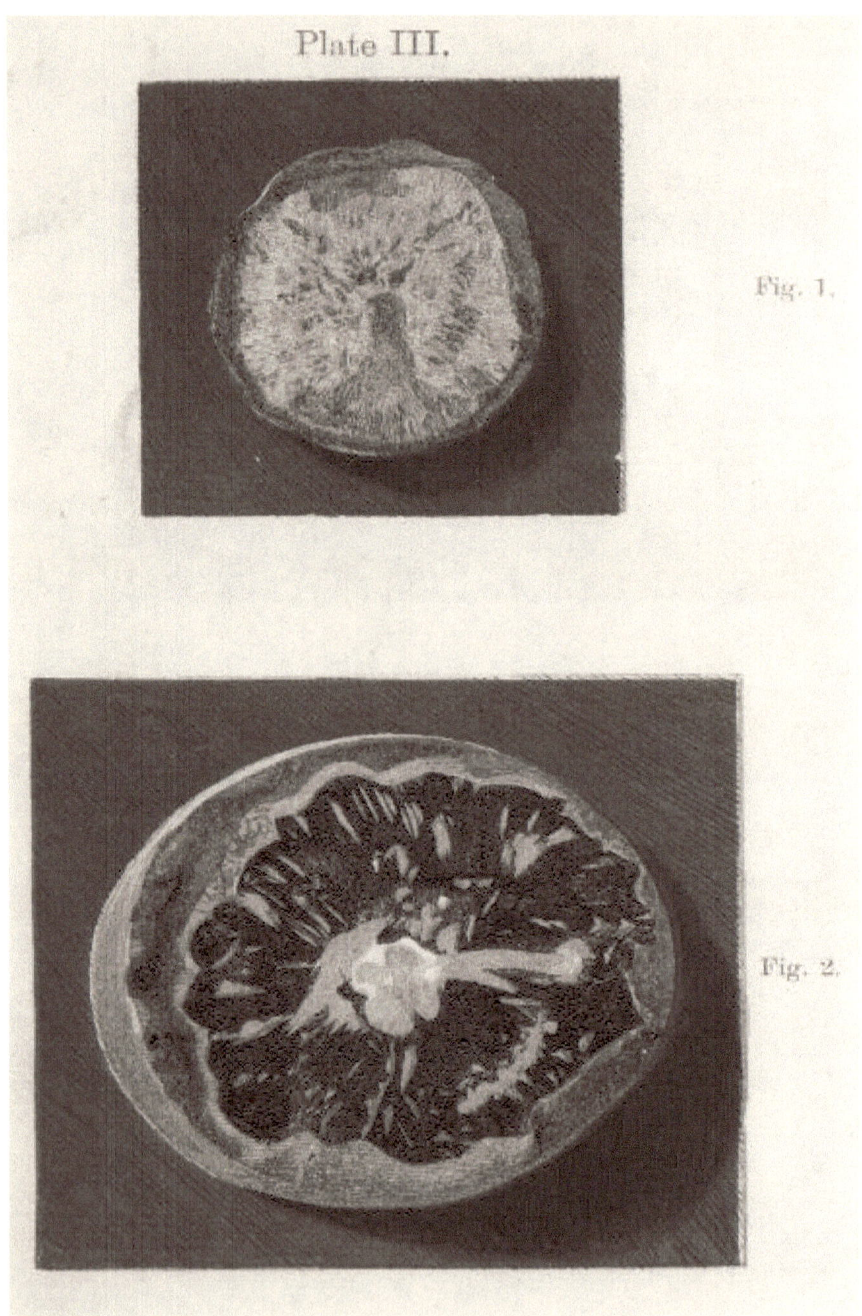

FIG. 1. FIG. 2.

PLATE IV.

Plate IV.

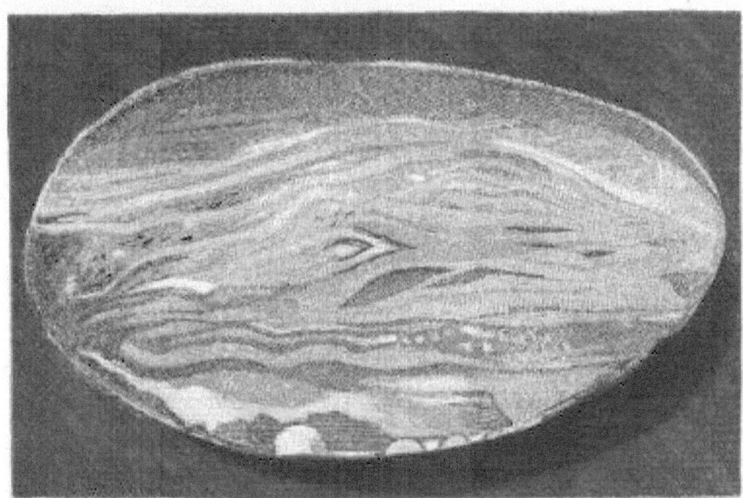

Fig. 1

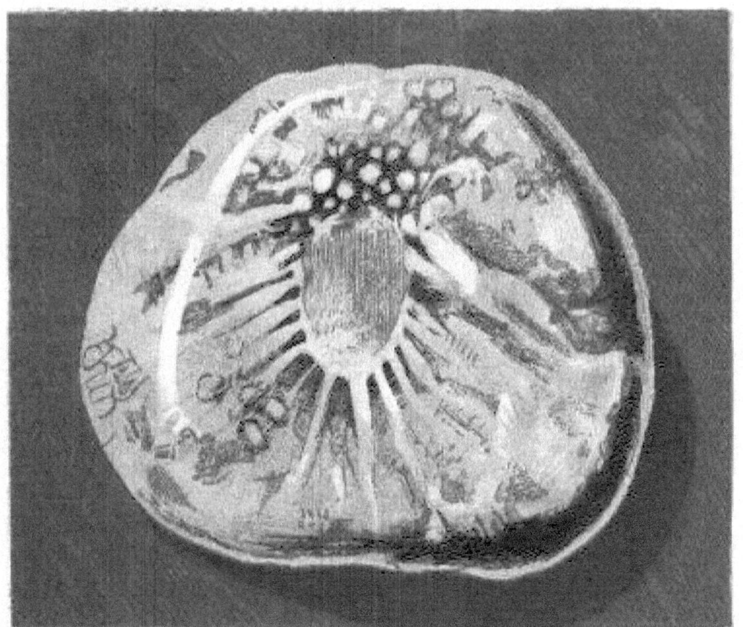

Fig. 2

FIG. 1. FIG. 2.

PLATE V.

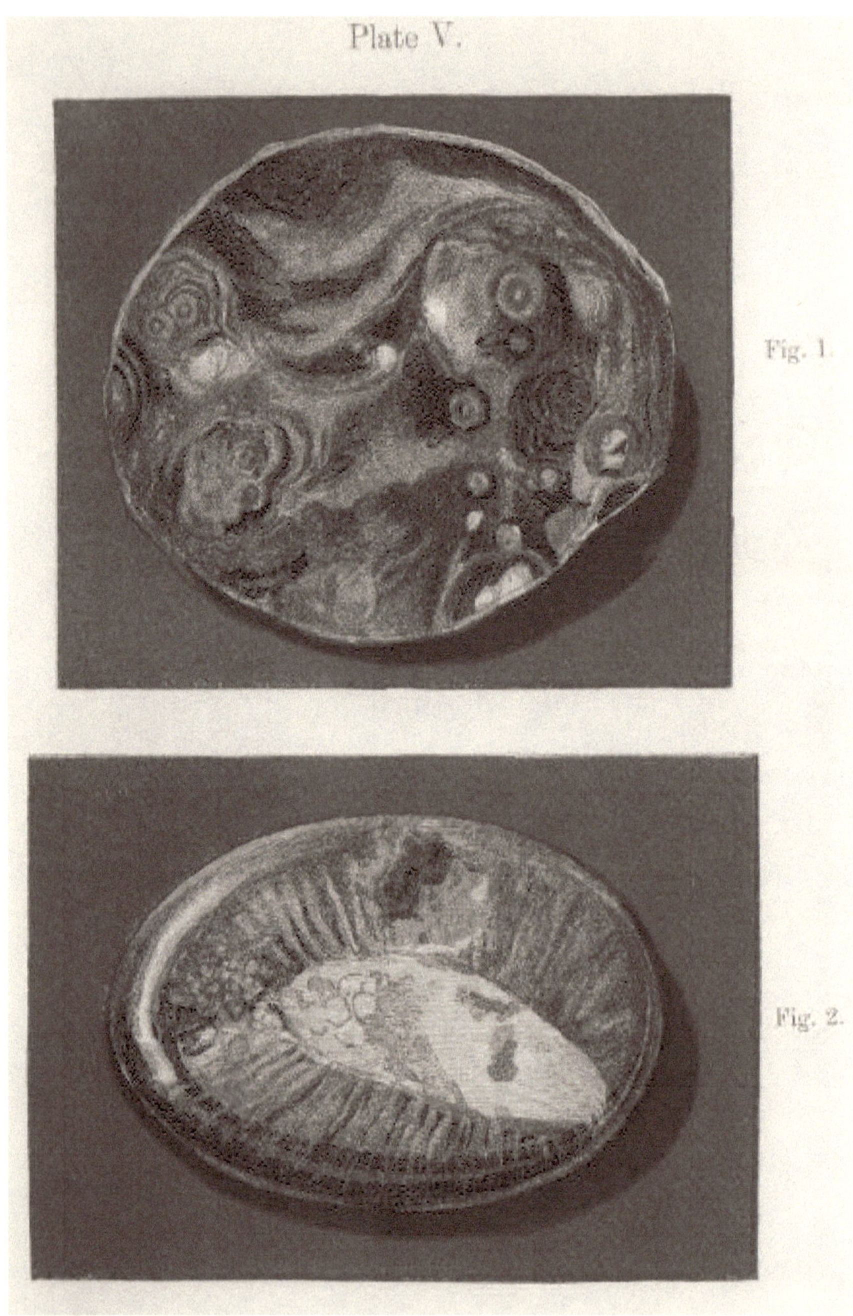

FIG. 1. FIG. 2.

PLATE VI.

FIG. 1. FIG. 2.

PLATE VII.

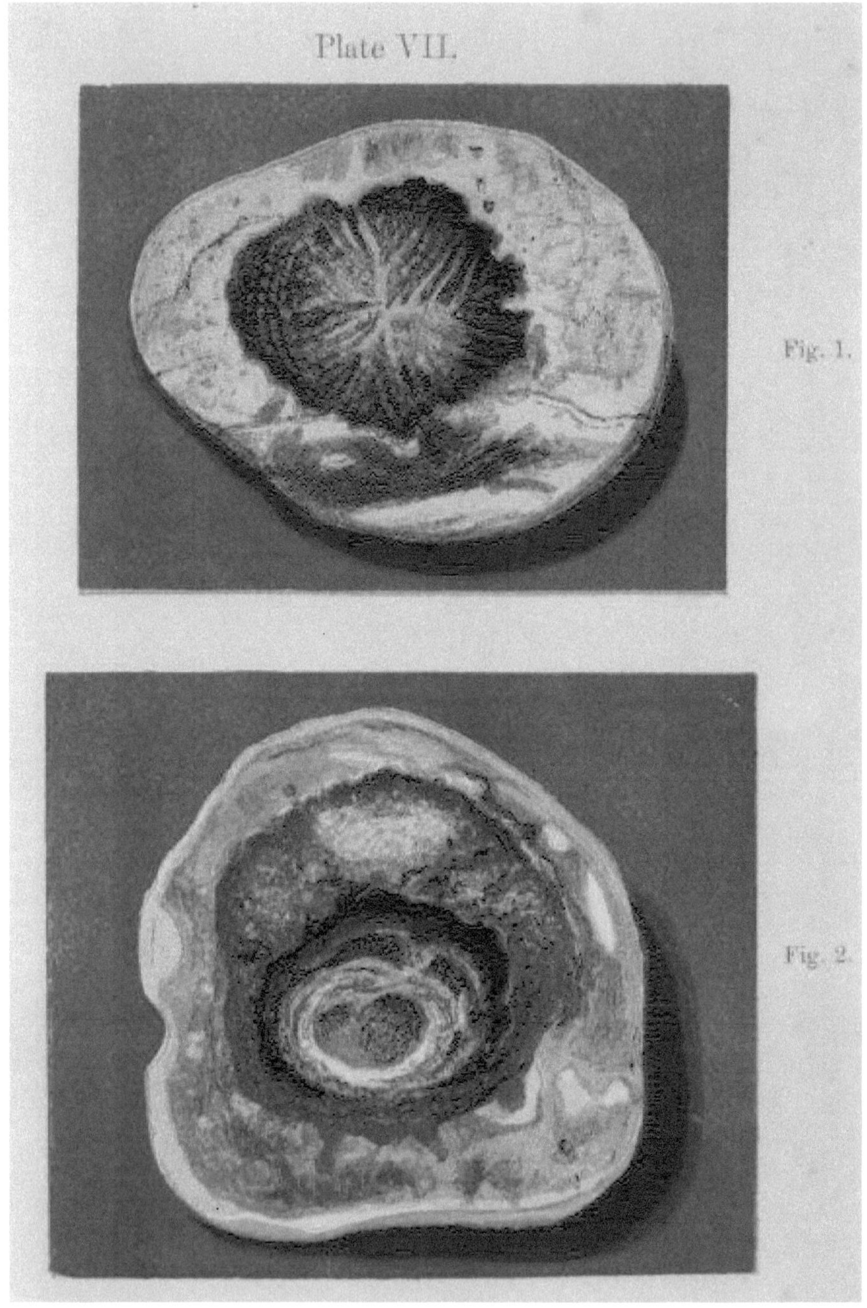

FIG. 1. FIG. 2.

INDEX

Lector House believes that a society develops through a two-fold approach of continuous learning and adaptation, which is derived from the study of classic literary works spread across the historic timeline of literature records. Therefore, we aim at reviving, repairing and redeveloping all those inaccessible or damaged but historically as well as culturally important literature across subjects so that the future generations may have an opportunity to study and learn from past works to embark upon a journey of creating a better future.

This book is a result of an effort made by Lector House towards making a contribution to the preservation and repair of original ancient works which might hold historical significance to the approach of continuous learning across subjects.

HAPPY READING & LEARNING!

LECTOR HOUSE LLP
E-MAIL: lectorpublishing@gmail.com